Lecture Notes
in Business Information Processing   146

Series Editors

Wil van der Aalst
Eindhoven Technical University, The Netherlands
John Mylopoulos
University of Trento, Italy
Michael Rosemann
Queensland University of Technology, Brisbane, Qld, Australia
Michael J. Shaw
University of Illinois, Urbana-Champaign, IL, USA
Clemens Szyperski
Microsoft Research, Redmond, WA, USA

W0230426

Henderik A. Proper
David Aveiro
Khaled Gaaloul (Eds.)

Advances in
Enterprise Engineering VII

Third Enterprise Engineering
Working Conference, EEWC 2013
Luxembourg, May 13-14, 2013
Proceedings

 Springer

Volume Editors

Henderik A. Proper
Public Research Centre - Henri Tudor
Luxembourg-Kirchberg, Luxembourg
E-mail: e.proper@acm.org

David Aveiro
University of Madeira
Funchal, Portugal
E-mail: daveiro@uma.pt

Khaled Gaaloul
Public Research Centre - Henri Tudor
Luxembourg-Kirchberg, Luxembourg
E-mail: khaled.gaaloul@tudor.lu

ISSN 1865-1348   e-ISSN 1865-1356
ISBN 978-3-642-38116-4   e-ISBN 978-3-642-38117-1
DOI 10.1007/978-3-642-38117-1
Springer Heidelberg Dordrecht London New York

Library of Congress Control Number: 2013936542

ACM Computing Classification (1998): J.1, H.3.5, H.4-5

Typesetting: Camera-ready by author, data conversion by Scientific Publishing Services, Chennai, India

Printed on acid-free paper

Springer is part of Springer Science+Business Media (www.springer.com)

Preface

Enterprise engineering is an emerging discipline that studies enterprises from an engineering perspective. Enterprises are studied as being purposely designed and implemented systems. Enterprise engineering is rooted in both the organizational sciences and the information system sciences. The rigorous integration of these traditionally disjoint scientific areas has become possible after the recognition that communication is a form of action. The operating principle of organizations is that actors enter into and comply with commitments, and in doing so bring about the business services of the enterprise. This important insight clarifies the view that that enterprises belong to the category of social systems, i.e., its active elements (actors) are social individuals (human beings). The unifying role of human beings makes it possible to address problems in a holistic way, to achieve unity and integration in bringing about any organizational change.

Also when regarding the implementation of organizations by means of modern information technology (IT), enterprise engineering offers innovative ideas. In a similar way as the ontological model of an organization is based on atomic elements (namely, communicative acts), there is an ontological model for IT applications. Such a model is based on a small set of atomic elements, such as data elements and action elements. By constructing software in this way, the combinatorial effects (i.e., the increasing effort it takes in the course of time to bring about a particular change) in software engineering can be avoided.

The development of enterprise engineering requires the active involvement of a variety of research institutes and a tight collaboration between them. This is achieved by a continuously expanding network of universities and other institutes, called the CIAO! Network (www.ciaonetwork.org). Since 2005 this network has organized the annual CIAO! Workshop, and since 2008 its proceedings have been published as *Advances in Enterprise Engineering* in the Springer LNBIP series. From 2011 on, this workshop was replaced by the Enterprise Engineering Working Conference (EEWC). This book contains the proceedings of the third EEWC, which was held in Luxembourg.

May 2013

Henderik A. Proper
David Aveiro
Khaled Gaaloul

Enterprise Engineering – The Manifesto

Introduction

This manifesto presents the focal topics and objectives of the emerging discipline of enterprise engineering, as it is currently theorized and developed within the CIAO! Network. There is close cooperation between the CIAO! Network (www.ciaonetwork.org) and the Enterprise Engineering Institute (www.ee-institute.com) for promoting the practical application of enterprise engineering. The manifesto comprises seven postulates, which collectively constitute the *enterprise engineering paradigm* (EEP).

Motivation

The vast majority of strategic initiatives fail, meaning that enterprises are unable to gain success from their strategy. Abundant research indicates that the key reason for strategic failures is the lack of coherence and consistency among the various components of an enterprise. At the same time, the need to operate as a unified and integrated whole is becoming increasingly important. These challenges are dominantly addressed from a functional or managerial perspective, as advocated by management and organization science. Such knowledge is necessary and sufficient for managing an enterprise, but it is inadequate for bringing about changes. To do that, one needs to take a constructional or engineering perspective. Both organizations and software systems are complex and prone to entropy. This means that in the course of time, the costs of bringing about similar changes increase in a way that is known as combinatorial explosion. Regarding (automated) information systems, this has been demonstrated; regarding organizations, it is still a conjecture. Entropy can be reduced and managed effectively through modular design based on atomic elements. The people in an enterprise are collectively responsible for the operation (including management) of the enterprise. In addition, they are collectively responsible for the evolution of the enterprise (adapting to needs for change). These responsibilities can only be borne if one has appropriate knowledge of the enterprise.

Mission

Addressing the afore-mentioned challenges requires a paradigm shift. It is the mission of the discipline of enterprise engineering to develop new, appropriate theories, models, methods and other artifacts for the analysis, design, implementation, and governance of enterprises by combining (relevant parts of) management and organization science, information systems science, and computer science. The ambition is to address (all) traditional topics in said disciplines from the enterprise engineering paradigm. The result of our efforts should be theoretically rigorous and practically relevant.

Postulates

Postulate 1

In order to perform optimally and to implement changes successfully, enterprises must operate as a unified and integrated whole. *Unity* and *integration* can only be achieved through *deliberate enterprise development* (comprising design, engineering, and implementation) and *governance*.

Postulate 2

Enterprises are essentially social systems, of which the elements are human beings in their role of *social individuals*, bestowed with appropriate *authority* and bearing the corresponding *responsibility*. The *operating principle* of enterprises is that these human beings enter into and comply with *commitments* regarding the products (services) that they create (deliver). Commitments are the results of *coordination acts*, which occur in universal patterns, called *transactions*.

Note. Human beings may be supported by technical artifacts of all kinds, notably by ICT systems. Therefore, enterprises are often referred to as sociotechnical systems. However, only human beings are responsible and accountable for what the supporting technical artifacts do.

Postulate 3

There are two distinct perspectives on enterprises (as on all systems): *function* and *construction*. All other perspectives are a subdivision of one of these. Accordingly, there are two distinct kinds of models: *black-box models* and *white-box models*. White-box models are *objective*; they regard the construction of a system. Black-box models are *subjective*; they regard a function of a system. *Function is not a system property* but a relationship between the system and some stakeholder(s). Both perspectives are needed for developing enterprises.

Note. For convenience sake, we talk about the business of an enterprise when taking the function perspective of the customer, and about its *organization* when taking the construction perspective.

Postulate 4

In order to manage the complexity of a system (and to reduce and manage its entropy), one must start the constructional design of the system with its *ontological model*. This is a fully implementation-independent model of the *construction* and the *operation* of the system. Moreover, an ontological model has a *modular* structure and its elements are (ontologically) *atomic*. For enterprises the meta-model of such models is called *enterprise ontology*. For information systems the meta model is called *information system ontology*.

Note. At any moment in the lifetime of a system, there is only one ontological model, capturing its actual construction, though abstracted from its implementation. The ontological model of a system is comprehensive and concise, and extremely stable.

Postulate 5

It is an *ethical necessity* for bestowing authorities on the people in an enterprise, and having them bear the corresponding responsibility, that these people are able to *internalize* the (relevant parts of the) *ontological model* of the enterprise, and to constantly validate the correspondence of the model with the operational reality.

Note. It is a duty of enterprise engineers to provide the means to the people in an enterprise to internalize its ontological model.

Postulate 6

To ensure that an enterprise operates in compliance with its *strategic concerns*, these concerns must be transformed into generic functional and constructional *normative principles*, which guide the (re-)development of the enterprise, in addition to the applicable specific requirements. A coherent, consistent, and hierarchically ordered set of such principles for a particular class of systems is called an *architecture*. The collective architectures of an enterprise are called its *enterprise architecture*.

Note. The term "architecture" is often used (also) for a model that is the outcome of a design process, during which some architecture is applied. We do not recommend this homonymous use of the word.

Postulate 7

For achieving and maintaining unity and integration in the (re-)development and operation of an enterprise, organizational measures are needed, collectively called *governance*. The *organizational competence* to take and apply these measures on a continuous basis is called *enterprise governance*.

May 2013   Jan L.G. Dietz

Organization

EEWC 2013 was the Third Working Conference resulting from a series of successful CIAO! Workshops over the years, the EEWC 2011 and the EEWC 2012. These events were aimed at addressing the challenges that modern and complex enterprises are facing in a rapidly changing world. The participants in these events share the belief that dealing with these challenges requires rigorous and scientific solutions, focusing on the design and engineering of enterprises.

This conviction led to the idea of annually organizing an international working conference on the topic of enterprise engineering, in order to bring together all stakeholders interested in making enterprise engineering a reality. This means that not only scientists are invited, but also practitioners. Next, it also means that the conference is aimed at active participation, discussion, and exchange of ideas in order to stimulate future cooperation among the participants. This makes EEWC a working conference contributing to the further development of enterprise engineering as a mature discipline.

The organization of EEWC 2013 and the peer review of the contributions to EEWC 2013 were accomplished by an outstanding international team of experts in the fields of enterprise engineering.

Advisory Board

Jan L.G. Dietz	Delft University of Technology, The Netherlands
Antonia Albani	University of St. Gallen, Switzerland

General Chair

Henderik A. Proper	Public Research Centre - Henri Tudor, Luxembourg Radboud University Nijmegen, The Netherlands

Program Chair

David Aveiro	University of Madeira, Madeira Interactive Technologies Institute and Center for Organizational Design and Engineering - INESC INOV Lisbon, Portugal

Organizing Chair

Khaled Gaaloul | Public Research Centre - Henri Tudor,
Luxembourg

Program Commitee

Bernhard Bauer	University of Augsburg, Germany
Birgit Hofreiter	Vienna University of Technology, Austria
Christian Huemer	Vienna University of Technology, Austria
Dai Senoo	Tokyo Institute of Technology, Japan
Eduard Babkin	Higher School of Economics, Nizhny Novgorod, Russia
Emmanuel Hostria	Rockwell Automation, USA
Eric Dubois	Public Research Centre - Henri Tudor, Luxembourg
Florian Matthes	Technical University of Munich, Germany
Gil Regev	École Polytechnique Fédérale de Lausanne (EPFL), Itecor, Switzerland
Graham McLeod	University of Cape Town, South Africa
Hans Mulder	University of Antwerp, Belgium
Jan Hoogervorst	Sogeti Netherlands, The Netherlands
Jan Verelst	University of Antwerp, Belgium
Joaquim Filipe	School of Technology of Setúbal, Portugal
Jorge Sanz	IBM Research at Almaden, California, USA
José Tribolet	INESC and Technical University of Lisbon, Portugal
Joseph Barjis	Delft University of Technology, The Netherlands
Junichi Iijima	Tokyo Institute of Technology, Japan
Marielba Zacarias	University of Algarve, Portugal
Martin Op 't Land	Capgemini, The Netherlands Antwerp Management School, Belgium
Natalia Aseeva	Higher School of Economics, Nizhny Novgorod, Russia
Olga Khvostova	Higher School of Economics, Nizhny Novgorod, Russia
Paul Johanesson	Stockholm University, Sweden
Peter Loos	University of Saarland, Germany
Pnina Soffer	Haifa University, Israel
Remigijus Gustas	Karlstad University, Sweden
Robert Lagerström	KTH - Royal Institute of Technology, Sweden
Robert Winter	University of St. Gallen, Switzerland

Rony Flatscher   Vienna University of Economics and Business
                                Administration, Austria
Sanetake Nagayoshi   Tokyo Institute of Technology, Japan
Stijn Hoppenbrouwers   HAN University of Applied Sciences,
                                The Netherlands
Ulrich Frank   University of Duisburg-Essen, Germany

Table of Contents

Value-Oriented Solution Development Process: Uncovering the Rationale behind Organization Components

João Pombinho[1,2], David Aveiro[3], and José Tribolet[1,2]

[1] CODE - Center for Organizational Design & Engineering, INESC INOV,
Rua Alves Redol 9, Lisbon, Portugal
[2] Department of Information Systems and Computer Science, Instituto Superior Técnico
Technical University of Lisbon, Portugal
[3] Exact Sciences and Engineering Centre, University of Madeira, Funchal, Madeira, Portugal
jpombinho@acm.org, daveiro@uma.pt, jose.tribolet@inesc.pt

Abstract. Although significant progresses have been made in recent years regarding the goals of Enterprise Engineering, we find that the rationale behind every component of an organization is still not systematically and clearly specified. Indeed, state of the art approaches to enterprise development processes do not explicitly incorporate an essential dimension of analysis: value. This state of affairs does not warrant a leading role in enterprise alignment.

We propose to address this issue by specifying a value-aware system development process and a system development organization. To this end, we began by applying DEMO to model the system development organization. Furthermore, the original Generic System Development Process (GSDP) was modelled, and improvement points identified. Our main contribution is a revision of the GSDP, combined with research on value modelling and enterprise architecture that explicitly includes the teleological part of the system development process.

The explicitation of the development process focusing on the value dimension, contributes to providing traceability and clarifying the rationale of each organizational artefact. We believe that modelling this rationale systematically will improve reactive and proactive change management through increased self-awareness, improved scenario specification, objective evaluation and well-grounded system development decisions.

Keywords: DEMO, GSDP, Value Modelling, e3Value, Solution Development.

1 Introduction

Business complexity and environmental change pace coupled with increasing ICT support exponentially increases the entropy of business systems. The mechanisms humans use to manage the complexity inherent to these systems and their dynamics pose various challenges, as they are not based on transversal, coherent and concise models. At the same time, cost reduction through effective reuse, reengineering and

H.A. Proper, D. Aveiro, and K. Gaaloul (Eds.): EEWC 2013, LNBIP 146, pp. 1–16, 2013.

innovation being heavily demanded features from enterprises and their supporting systems. Laudon notes that enterprise performance *is optimized when both technology and the organization mutually adjust to one another until a satisfactory fit is obtained* [1]. However, studies indicate as much as 90 percent of organizations fail to succeed in applying their strategies [2].

Misalignment between the *business* and its *support systems* is frequently appointed as a reason of these failures [1, 3]. Aligning Business and IT is a widely known challenge in enterprises as the developer of a system is mostly concerned with its function and construction, while its sponsor is concerned about its purpose, i.e., the system's contribution. Also, the business vision of a system and its implementation by supporting systems is not modelled in a way that adequately supports the development and evolution of a system and its positioning in a value network. A paradigm shift in the way of modelling and developing systems must occur so that they can be increasingly developed considering their dynamic context and formally addressing the rationale behind value network establishment and system/subsystem bonding.

Formally integrating the notion of purpose into system development activities requires addressing both the teleological and ontological perspectives in an integrated, bidirectional way [4]. However, Engineering approaches are generally focused solely on the ontological perspective [5]. By Enterprise Engineering is meant the whole body of knowledge regarding the development, implementation, and operation of enterprises [6]. DEMO has a particularly relevant role in this area both as ontology and as a method. The Generic System Development Process (GSDP) is specified in DEMO's TAO-theory as the process by which a system is designed and implemented from the specifications of its using systems. The GSDP is systematically defined, clarifying normally ambiguous concepts like architecture, design, engineering and implementation. However, it lacks in instantiation and practical demonstration of usefulness.

This paper addresses the mentioned challenges by combining enterprise engineering and value modelling and is structured as follows: section 2 presents related work and the problem at hand. Section 3 introduces a practical scenario that will be used for reference through the paper. In section 4, we present our solution proposal and a more detailed instantiation of the method, with localized analysis. The paper closes with contribution summary and conclusions.

2   Related Work and Problem Statement

2.1   Related Work

In this section we introduce the enterprise engineering (EE) discipline and enterprise ontology and DEMO, a theory and method of EE. Next, we present e3Value, an approach to value modelling.

2.1.1   Enterprise Ontology and DEMO

Enterprise ontology [6] includes a sound theory and a method for supporting enterprise engineering. It goes beyond traditional function (black-box) perspective aiming at changing organizations based on the construction (white-box) perspective.

Organizations are considered as systems composed of social actors and their interactions in terms of social commitments regarding the production of business facts.

From the Transaction Axiom of Enterprise Ontology, we find that actors perform two kinds of acts. By performing production acts (P-acts), the actors contribute to bringing about and delivering services to the environment. By performing coordination acts (C-acts), actors enter into and comply with commitments. An actor role is defined as a particular, atomic 'amount' of authority, viz. the authority needed to perform precisely one kind of production act. P-acts and C-acts occur in generic recurrent patterns, called transactions. Every transaction process is some path through this complete pattern, and every business process in every organization is a connected collection of such transaction processes [6].

From the Distinction Axiom of Enterprise Ontology's PSI-theory, we find that we can divide all acts of an organization in 3 categories - ontological, infological and datalogical, respectively related with the 3 human abilities: performa (deciding, judging, etc.), informa (deducing, reasoning, computing, etc.) and forma (storing, transmitting, etc.). By applying both axioms, Enterprise Ontology's Design and Engineering Methodology for Organizations (DEMO) is able to produce concise, coherent and complete models with a dramatic reduction of complexity.

Unlike other approaches, DEMO makes a very strict distinction between teleology, concerning system function and behaviour – the black-box perspective – and ontology, about its construction and operation – the white-box perspective [7]. These perspectives are embodied in the Generic System Development Process (GSDP), represented in Figure 1. It begins with the need by a system, the Using System (US), of a supporting system, called the Object System (OS).

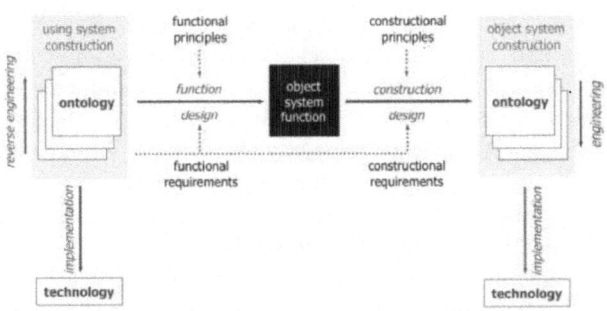

Fig. 1. Generic System Development Process [6]

From the white-box model of the US, one determines the functional requirements for the OS (function design), formulated in terms of the construction and operation of the US. Next, specifications for the construction and operation of the OS are devised, in terms of a white-box model (construction design). The US may also provide constructional (non-functional) requirements. Choices are then made with each transition from the top-level white-box model towards the implementation model. However, nothing is prescribed about the rationale behind these choices. System design decisions, either implicit or explicit, remain solely, and certainly not forever, in the minds of the participants in the process. The sheer complexity can quickly cross the limits of unsupported human handling. It may then become short of impossible to know the rationale of past decisions, its impacts and dependencies in designing the to-be.

2.1.2 Developing Organizations with Control and G.O.D Sub-organizations

Aveiro took a step towards instantiating the GSDP by applying DEMO to specify the models of the sub-organizations responsible for handling change caused by exceptions. In the control sub-organization [8], the viability of a system is specified by a set of measures and respective viability norms that can be periodically checked against the operational status. If such norms are violated, a dysfunction handling mechanism is triggered. If the exception that causes the dysfunction to the norm is expected, solutions that have previously been identified in anticipation are applied and evaluated for solving the problem. If the cause is unexpected, an organizational engineering process (OEP) must be started, that occurs in the scope of another sub-organization, the G.O.D. organization [9], responsible for specifying and implementing change that will solve or circumvent the unexpected exception causing a dysfunction. The solution may be new organizational components (e.g., new norms, new actors, processes and rules, etc.) or just (re-)allocation of human or IT resources.

2.1.3 Value Modelling – e3Value

There are many classifications of organizations, according to their composition and objectives, including: private, public, political, business, educational, healthcare, non-profit, etc. All organizations have in common bringing about *value* to their *environment*, either directly or indirectly, so *value* is an unifying concept. Also, Value Modelling was selected as it is increasingly recognized that the concept of value assists in improving stakeholder communication, particularly Business and IT [10].

e3Value [11] is part of e3family, a set of ontological approaches for modelling networked value constellations. It is directed towards e-commerce and analyses the creation, exchange and consumption of economically valuable objects in a multi-actor network. In e3Value, an Actor is perceived by his or her environment as an economically independent entity, exchanging Value Objects. An enterprise is modelled as an actor in a value network, where the demand and offer market concepts are a natural consequence of the economic context of Value Objects.

As will be presented in section 4, we propose applying e3Value to improve system and subsystem value modelling: inside the boundaries of organizations, as opposed to applying it solely to e-commerce relations between formal organizations.

2.2   Problem Statement

Looking at previous efforts on formalizing organization development, one question that comes to mind is: what are the criteria for generating new organizational components? In [12] generic acts of monitoring, diagnosis and recovery are used to specify the rationale behind change. But such categorization is quite generic and does not explicitly capture an essential dimension of analysis: value. As an example, we can think of a viability norm as the minimal number of movie loans per month at a video store. In practice, this is an economic condition for having minimal profit required for sustained survival and growth of Watch-it business, the generic and main value condition for the company. However, if only a "local" perspective is taken during viability norms specification, global, combined effects of these and other value drivers are missed. Still, broader rules can be applied and the combined effect of drivers can be calculated and set as a wider rule. But even so, the very structure of the organization and the reasoning behind these rules may not be precisely captured.

We hypothesize that these rules are set during the implementation of not only reactive (the focus of GOD) but also proactive and evolutive changes of the organization. Such rules must not only conform to but justify its structure as there is a bidirectional relationship between value conditions, value network and the organizational structure as well as the resources needed to "implement and run it".

During a system development process, the designed system/subsystem relations are a result of choices between different solutions for intermediate and possibly interconnected sub-problems. Such sub-solutions can and should be modelled as individual system development efforts, preserving the modularity that allows for rigorous modelling and tracing of the rationale behind these intermediate choices. By defining a formal model of the development process, the relations between systems and sub-systems can be made explicit as problem/solution pairs, thereby explicitating the nature of these relations and flattening the system structure, while preserving rational structure as it will be explained in section 4.2.

In order to clarify our solution proposal to these issues, we chose to model the system development organization. It seemed appropriate to apply DEMO to the GSDP itself, as a system development organization, and defining its own ontological model. The results were then combined with previous research on value modelling [4, 13].

In the following sections we explore the reasoning just presented and research results in two phases. The first, intended as a formalization of the GSDP as defined in [6]. The second phase is a revision of the GSDP to include the teleological part of a given system development process.

3 Unimedia Case: Remote Internet Customer Support

Unimedia is a quadruple-play operator (television, internet, fixed voice and mobile voice) with a large customer base. Customers may have a combination of services and some services require customer premises equipments (CPE). These equipments amount to a relevant part of customer support, particularly for the internet service. The remote customer support organization is described by the following narrative:

In the case there is a perceived malfunction by the customer, she can contact the call center directly to identify and solve the issue. After calling the support number, her call is handled by an Interactive Voice Response (IVR) system. IVR allows customers to interact with the company via telephone keypad or by speech, so they can service their own inquiries by following the predefined process or, eventually, get redirected to a human operator. The client identifies by dialling the national ID number. Additional identification information can be requested for cross-check later in the call if there are relevant actions to be taken. Following, a diagnosis process is carried on. The diagnosis can be at customer side (e.g. check the modem lights) or at the provider side (e.g. check service provisioning status). After a diagnosis is established, a solution is attempted. Again, the solution can be at the customer side (e.g. reset device) or at the provider side (e.g. force firmware update). The call ends after reaching a solution or, if it is not successful, by requesting field service.

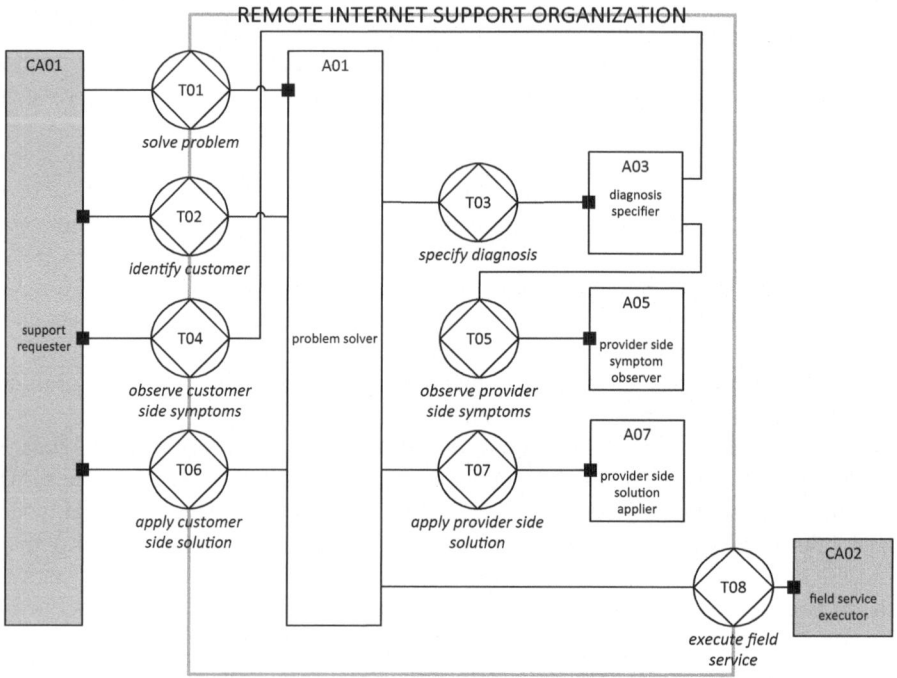

Fig. 2. Unimedia Remote Internet Support Actor Transaction Diagram (ATD)

Following the alignment process described in [14], an extension to the Transaction Result Table (TRT) was proposed, including the concepts of Value Object (VO) and Value Transaction. The resulting value model is shown in Figure 3.

Table 1. TRT extended with Value Object and Value Transaction

Transaction kind	Transaction result	Value Object	Value Transaction
T01 solve problem	R01 problem P has been solved	P solution	VT01 Problem solution for Money (contract)
T02 identify customer	R02 customer contract C has been identified	C contract	VT01 Problem solution for Money (contract)
T03 specify diagnosis	R03 diagnosis D has been specified	diagnosis	VT02 Diagnosis for Money
T04 observe customer side symptoms	R04 customer side symptom CSS has been observed	CS symptom	VT03 CSS for Eligibility
T05 observe provider side symptom	R05 provider side symptom PSS has been observed	PS symptom	VT04 PSS for Money
T06 apply customer side solution	R06 customer side solution CSP has been applied	CS action	VT05 CSP for Eligibility
T07 apply provider side solution	R07 provider side solution PSP has been applied	PS action	VT06 PSP for Money
T08 execute field service	R08 field service FS has been executed	FS solution	VT07 FS for Money

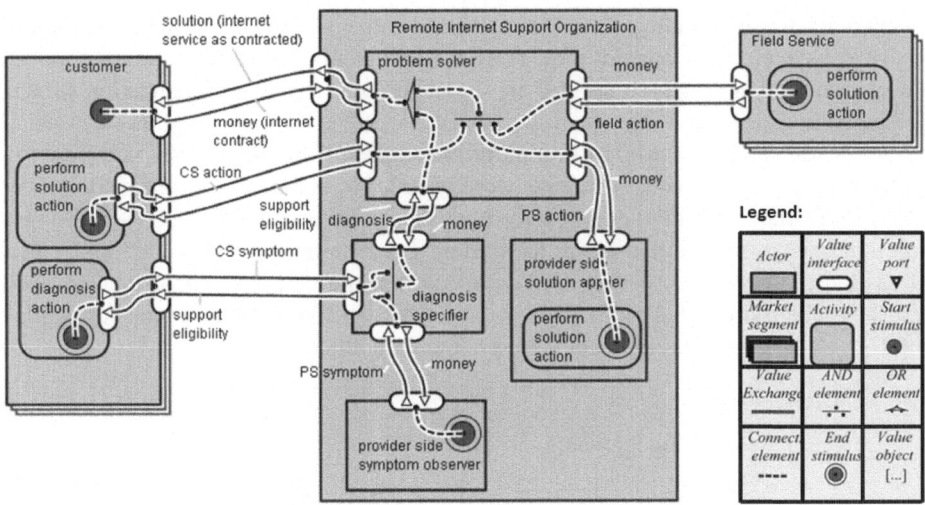

Fig. 3. Value model for Remote Internet Support scenario

The description of the process and benefits of aligning value and ontological models exceeds the scope of this paper and is presented in [13, 14]. Still, a brief example of contributions from both sides follows.

Some clarifications resulting from aligning ontology models based on value were the explicitation of value activities. For instance, as part of getting free remote customer support, the customer must provide "eyes & hands" to get *support eligibility*, which is the VO. Actually this is company policy but was missing from the narrative and was identified due to the notion of *economic reciprocity* from e3value – the transactions must have at least an inbound value port and an outbound value port. Also, note that *CS symptom* and *CS action* are relevant VOs because they are intermediate results for their respective solution chains: diagnose problem and solve problem.

On the opposite direction, the main contribution of ontological analysis is that social interaction theory and, particularly, the transactional pattern allow checking the value model for completeness and consistency. One example is testing the value object exchange over the complete transactional pattern, with possible impacts on (re-) specification of value objects and interfaces, e.g., what happens if a customer declines performance of local diagnosis?

4   Improving the GSDP - Introducing Purpose and Value

4.1   Applying DEMO Methodology to the GSDP

We define a solution to a problem as the production of a determined result, which generally involves investment of resources (time, money, effort, etc.) by the Object System (OS) and generates value for some stakeholder, the Using System (US). By asking the solution requester to define the construction of the US and its value model, additional insight can be derived from its specification. This insight can change the

problem or dissolve it altogether. However, the entry point of the GSDP, i.e., the origin of the system development request, is not sufficiently clear in the original model.
To overcome this issue, we defined the Solution Development Organization (SDO),
presented in Figure 4.

In our view, the description originally provided for the GSDP was not ontologically complete and some adjustments were in order to obtain a coherent model of the
SDO. Particularly, we defined a recurrent *provide solution* transaction (N+1) as a new
solution development cycle where the current OS assumes the role of US and a new
OS is being developed so that its function serves the construction of the US. This
transaction is represented by the link between A03 and T01 and is crucial for explicit
multi-cycle solution development, i.e., function/construction alternation.

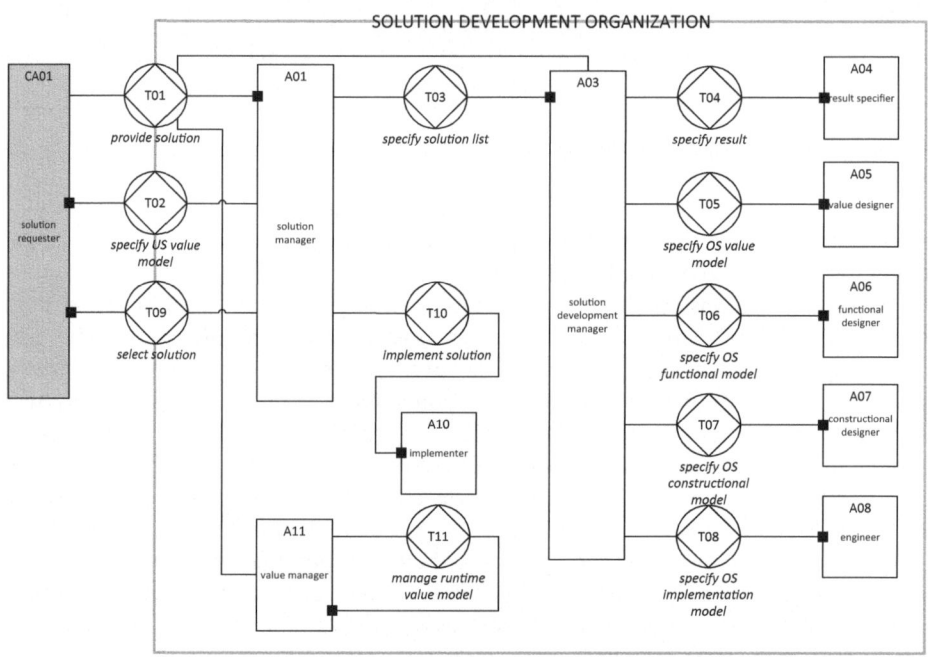

Fig. 4. Solution Development Organization – ATD

The process begins with an external request to provide a solution. In this case, Unimedia's Head of Customer Support requests a solution for reducing costs, following
a decision by the board that their internet support costs are to be reduced by at least
20%. The solution manager asks the requester to specify the Using System value
model, which is critical to identify rational solutions. In this case, the requester
produces a value model, showing that the largest costs come from the calls that get
redirected to human operators. The solution manager then requests that the solution
development manager specifies a solution list to produce the result requested, in the
context of the US value model. The specify results transaction is the creative step of

this process, where different ways of producing the required result (solution) are identified. For instance, one idea would be to recruit cheaper operators; another would be
that the less calls were redirected to the human operators. For each result, the value
model and functional model are specified in sequence. Next, the constructional model
is built, where transactions and actors are specified. In this case, the result would be to
lower the number of redirections to expensive human operators by 20%.

If there is a dependency in producing the result, then another solution development
process is triggered, with the solution development manager requesting a solution for
that problem. The current OS is repositioned, assuming the role of US in the new
development cycle. For instance, the dependency can be to find a solution to provide
additional checks and redirections to avoid costly human operators whenever possible. Such a request would be made by the level 1 solution provider to level 2 providers. For each crossing of these levels, a new GSDP iteration takes place. Along each
single thread of a solution chain, the alternation between each pair of levels is described by Dietz and Hoogervorst as function/construction alternation [6]. A set of
such iterations is commonly performed implicitly inside a single GSDP, and thereby
kept from being adequately modelled by the explicit application of functional,
constructional and architectural principles.

When the set of known solutions is considered satisfying by the solution manager,
it requires that the solution requester elects a solution from the presented alternatives.
The elected solution is implemented and its value proposal is periodically monitored
by the value manager. If an inconsistency is found, the *provide solution* transaction is
invoked to address the gap, presented as an economic viability problem.

4.2   The Method at Work: Value-Driven Cost Reduction

We now present the method inherent to the solution development organization. This
generic method applies to both a bootstrapping setting or to an ongoing change.

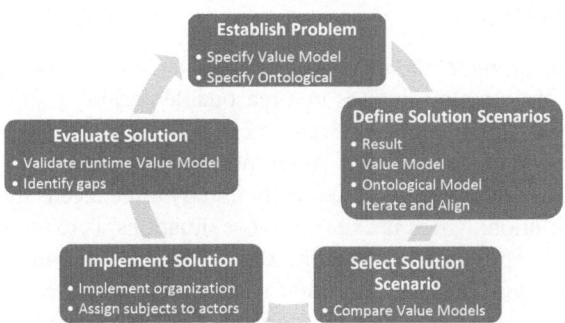

Fig. 5. VoSDP - Method for practical application

I - Establish Problem

Revisiting our example, let us begin with the initial request. The fact that the investors are the requester means they must come into play explicitly in the value model. The first step is to represent the *as-is* set. Due to space limitations, a simplified generic value model of a private, for-profit enterprise is presented in Figure 6.

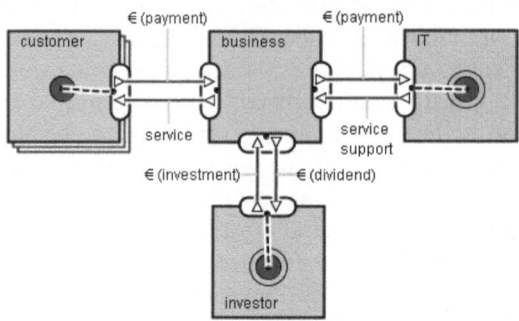

Fig. 6. Generic value model of IT-enabled for-profit enterprise

Considering the value model in Figure 6, the request can be reformulated as improving the investor balance (the real sought-after result). In this simple model, the equation for an annual period is:

$$dividend = customer\ revenue - business\ OPEX - IT\ OPEX - investment$$

The result can be attained by reducing the expenditure or finding alternative ways of generating value, such as increasing revenue (relating with the customer actor) or decreasing support costs (relating with the IT actor). Somewhat surprisingly, as we are about to see, the choice was increasing investment and IT OPEX costs.

II - Define Solution Scenarios

After clarifying the problem, the solution manager starts a solution development cycle that returns a list of possible solutions in a reasonable period of time. Please note that modelling the value of the solution development process itself and, therefore, obtaining a consolidated value model that takes into account return on modelling effort (ROME) can be done by using the same methodology but exceeds this paper's scope.

One obvious solution, which is exists in most situations, is to leave everything as it is. By default, this represents the baseline scenario. The solution development manager is to identify additional scenarios and begins as a mostly creative endeavour of identifying results/value objects that make up the following nodes on the value chain for obtaining the original result. In our example, it is necessary to know the cost structure of the *business* actor from Figure 6. For simplicity and space economy, let us consider that Figure 2 is a complete model of the business actor. In this case, it turns out that the *problem solver* stands out by a large margin while analyzing the transactional costs of the actors (part of the e3Value model). The fact that the ontological model of the organization does not allow concluding this is no surprise as it abstracts implementation. On the other hand, the value model also has an invariant perspective but it is complemented with selected implementation-level constructs. In this perspective, it is possible to include parameters that are implementation-level value estimates.

Returning to the example, the transactional costs of the problem solver are mostly due to the time the human agent spends in 1) initial call handling, i.e., identifying the customer and 2) filtering away exceptions to normal diagnosis.

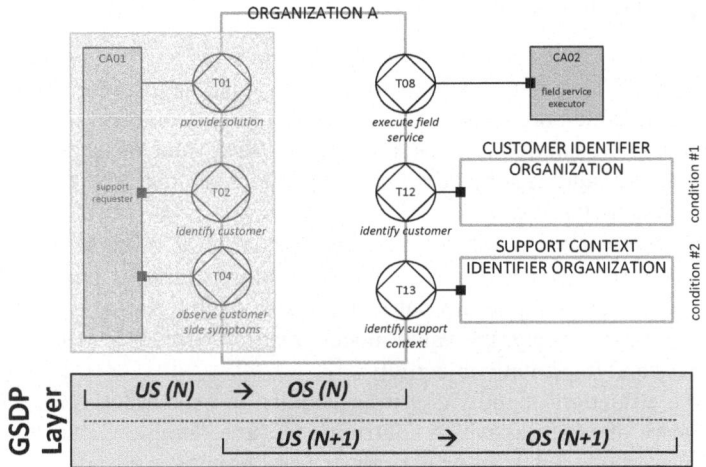

Fig. 7. Recursive SDP cycles for addressing identified problems

As represented in Figure 7, two (layer 2) system development processes (SDP) are started by the *solution development manager* (layer 1) with the request of reducing human operator time by each of the conditions mentioned above. The value model is already clarified, as it previously existed at an adequate detail level to specify the problem scope in a way that can be related to the ontological model.

The problem related to condition 1 can be solved by using existing CRM services to identify a customer by a set of keyed-in data that is sent to the IVR system via Dual-Tone Multi-Frequency (DMTF).

The problem related to condition 2 can be solved by using existing services that, based on the customer address and portfolio, can identify situations that constitute exceptions to the scope of the remote internet support system. For instance, Service Management ticketing system services can identify if there is a common problem in the geographical area of the customer service address/cell, e.g., a power blackout or a cut backbone. In these cases, there is no added value in handling the call to a human operator, so an automated vocal message that describes the situation and provides expected resolution time allows ending the call without resorting to a human agent.

Note that, in the case of condition 1, the *identify customer* transaction already exists and, despite being executed by the customer it is still problematic. The issue is in the complexity of the accept phase of the transaction pattern. By splitting the initiating actor of the transaction, it is now possible to allocate different subjects that can execute more efficiently. It can be argued that these solutions are simple automations and do not change the construction, being solely implementation choices. We are aware that these conditions are of infological nature but to implement each solution, it is necessary to add an actor to the construction. This actor has its own business with service levels and responsibilities, which is the same as saying we are dealing with a US and OS both at the B-level so it is a matter of relativity, as discussed in [4].

III - Select and Implement

In order to rationally select solutions scenarios, objective criteria must be defined. To this end, using e3Value it is possible to assign valuation formulas to value object transfers through value ports. There are two types of value objects: 1) money objects, when the amount transferred can be objectively stated and observed; and 2) non-money objects: the value is subjective, meaning actors can disagree about the amount of economic units they assign it.

While non-money objects can be important for design and impacts analysis, there should be an effort to monetize costs/benefits to allow financial analysis. The techniques and theory for doing so are out of the scope of this paper. Nevertheless, it is worth noting that, seemingly, the value model creation by itself is a step forward creating consensus and improving objectivity.

Besides the valuation of individual transactions on each value port, e3Value defines the concept of expenses which contribute to the economic viability analysis but are not explicitly modelled as value exchanges (e.g., employee costs).

- Variable expenses – occur multiple times per value model, depending on the transaction volume and are assigned to value ports, being useful for modelling operational expenditure (OPEX);
- Fixed expenses – occur only once per period, e.g. monthly wages, used for simplification, also useful for modelling OPEX;
- Investments – a particular fixed expense, occurring only once per time series (scenario) and therefore useful for modelling capital expenditure (CAPEX).

e3Value allows specifying value model components using specific attributes that make the profitability sheets directly derivable from the model. Table 2 and Table 3 represent simplified annual profitability sheets for both scenarios, where value port details have been excluded.

Table 2. Profitability sheet for scenario A

Actor	Interface	Occurences	Valuation	Value
problem solver	problem solution, internet contract	1080000	0,30 €	- 324.000 €
	EXPENSES			5.000 €
	INVESTMENT			- €
actor total				- 329.000 €
overall				- 329.000 €

Table 3. Profitability sheet for scenario B

Actor	Interface	Occurences	Valuation	Value
problem solver	problem solution, internet contract	918000	0,24 €	- 220.320 €
	EXPENSES			- 5.000 €
	INVESTMENT			- €
actor total				- 225.320 €
customer identifier	problem solution, internet contract	1080000	0,02 €	- 21.600 €
	EXPENSES			- 1.200 €
	INVESTMENT			- 15.000 €
actor total				- 37.800 €
support context identifier	problem solution, internet contract	1080000	0,01 €	- 10.800 €
	EXPENSES			- 1.200 €
	INVESTMENT			- 15.000 €
actor total				- 27.000 €
overall				- 290.120 €

For the sake of illustration of the concepts of this paper, we chose to represent only the cost stream, assuming stable customer revenue. Out of the 1,5M customers, based on historical data we considered that 900K have internet services and they call for support 1,2 times/year on average. We assume the implementation of the new actors has a CAPEX of 30 K€ and OPEX of 2,4 K€, allocated in halves to each. After implementation, the number of calls handled by problem solver has reduced by 15% and the average support call duration for this actor lowered from 10 to 8 minutes because of effectively reduced support scope due to early context clarification.

In this particular situation, the option is clear after comparison of the value models. Selecting scenario B is relatively straightforward, as there is an interesting business case versus scenario A. Scenario B represents an investment with a payback of 6 months and, onwards, a positive impact of 5,7 K€/month versus scenario A. It is noteworthy that the existence and relevance of an *investor* actor is formally required because it is included in overall value model. Obviously, there may be different scenarios and additional analysis with impact on their definition may be called for, resulting in additional iterations. For instance, the investor may try to find better solutions to invest his money and get payback in less than 6 months.

In this example we left out using time series, a concept that directly addresses the time variable and establishes value models for specific consecutive time periods. This view is useful not only for business case specification over time but also to align expected value production with solution architecture and construction roadmaps.

The relevant aspect of implementation we want to make clear, besides its technicalities, is that the implementation of the artefacts is also accompanied by putting the business model itself into operation (production environment). We refer to this as a *live business model*, in opposition of using it solely for evaluation and decision purposes early in the process. This means that the value model is now an artefact which is controlled by a specific actor, *value manager*. The *value manager* compares operational reality to the specification in the model and may decide to request the problem solving organization to address a potential gap. While the detailed specification of how this comparison is carried through is out of the scope of this article, it is relevant to note that it is enabled by the existence of specific constructs to model value.

IV - Evaluation

Evaluation happens both at the implementation review of a project and continuously at runtime, in the spirit of the *live business model* concept. The exhaustive description and analysis of this phase exceeds the scope of this paper, but it can be concluded that the explicitation of the development process and the intermediate deliverables produced contribute to the availability and objectivity of evaluation mechanisms. In our example, if the implementation of the services that provide information for identifying support context is consistently unreliable, the projected benefits may not be achieved and may even have negative results because of mistaken call redirections. The advantage of having a business case integrated with the ontological and implementation models is that it is now possible to estimate the critical values that put economic viability at stake and monitor them in anticipation via trend analysis.

Evaluation can also lead to exploring alternative ways that were not selected but that were considered at an earlier phase of the solution development process. Leveraging the prospective solutions concept presented earlier lets us, e.g., return to the original solution request and the idea of increasing revenue. One way of contributing to this value stream is by reusing the automated IVR-based solution just developed as a channel for up selling/cross-selling. This repositions the customer care organization from a cost center to a value center. The opportunity of having the customer in-line can be taken advantage of by creating a discounted offer for these situations, to be presented automatically (relatively inexpensive) and/or redirect the call to a sales operator (more costly; more effective?). To explore this path, a new GSDP cycle is in sight. Only after successful solving the customer problem, of course!

5   Conclusion

We found that in order to capture the rationale behind organizational artefacts, we need additional constructs to those DEMO currently provides. The contributions of this paper can be summarized as redesigning the GSDP and the corresponding SDO for supporting multiple cycles and extending it with value concepts. Alternating Value/Function/Construction in successive cycles was found relevant and applicable.

As explained on section 3, the main contribution of ontological analysis to the match with value modelling is that social interaction theory and, particularly, the transactional pattern, allows checking the value model for completeness and consistency. Conversely, by integrating value modelling with ontological modelling we can anticipate decisions based on projected implementation viability and leave a formal trace of the decision rationale. Moreover, we exemplified how the resources used in the implementation of the system may relevantly restrict the ontology of the system: 1) there are ontological subsystems purely constructed by some value condition and 2) the value specification must be part of the production world. Very frequently, parts of the construction depicted in the ontological model depend on value constraints at implementation level and to strive for fully implementation independent models would be either unfeasible or a simplistic approach with unuseful models as a result. Therefore, we see ontological models not as implementation *independent*, but rather implementation *abstracted*.

The benefits from these contributions go beyond the simple support system automation and rest on the capacity to model the essentials of the businesses involved in their commonalities and differentiators. Each variation point of a business area places demands on the construction of the organization providing these services. These services are valued distinctly by different customer types and this value should be actively managed in articulation with the construction. In turn, they allow exploring synergies through reutilization of solutions and increased insight given by explicitating the intermediate artefacts of the solution development process.

For all scenarios considered, even if the solution development step is not completed for some reason, e.g., lack of investment capability or analysis time, every deliverable is kept in association with the problem specification. While this is somehow obvious for complex deliverables, such as value models, even a simple enumeration of results in a hypothetical chain, with generalization or specialization of the value objects, represents prospective solutions that can be revisited later on.

As it can be seen from the example, there is no magic bullet regarding creative solution hypothesizing. As a practical observation and clarification, our method allows domains experts to be involved by the responsible actors in both the solution development and selection transactions. Some mechanisms based on knowledge about prospective or used solutions, for instance generalization/specialization of value objects may be used as a starting point. Still, there is no greater ambition than to provide useful tools for the human mind to do its job.

References

1. Laudon, K.C., Laudon, J.P.: Management Information Systems: Managing the Digital Firm. Prentice Hall (2011)
2. Kaplan, R.S., Norton, D.P.: Strategy Maps: Converting Intangible Assets Into Tangible Outcomes. Harvard Business School Press, Boston (2004)
3. Henderson, J.C., Venkatraman, N.: Strategic alignment: leveraging information technology for transforming organizations 32(1), 4–16 (1993)
4. Pombinho, J., Aveiro, D., Tribolet, J.: Towards Objective Business Modeling in Enterprise Engineering – Defining Function, Value and Purpose. In: Albani, A., Aveiro, D., Barjis, J. (eds.) EEWC 2012. LNBIP, vol. 110, pp. 93–107. Springer, Heidelberg (2012)
5. Op 't Land, M., Pombinho, J.: Strengthening the Foundations Underlying the Enterprise Engineering Manifesto. In: Albani, A., Aveiro, D., Barjis, J. (eds.) EEWC 2012. LNBIP, vol. 110, pp. 1–14. Springer, Heidelberg (2012)
6. Dietz, J.L.G.: Enterprise Ontology: Theory and Methodology. Springer (2006)
7. Dietz, J.L.G.: Architecture - Building strategy into design. Netherlands Architecture Forum, Academic Service - SDU, The Hague, The Netherlands (2008)
8. Aveiro, D., Silva, A.R., Tribolet, J.: Control Organization: A DEMO Based Specification and Extension. In: Albani, A., Dietz, J.L.G., Verelst, J. (eds.) EEWC 2011. LNBIP, vol. 79, pp. 16–30. Springer, Heidelberg (2011)

9. Aveiro, D., Silva, A.R., Tribolet, J.: Extending the Design and Engineering Methodology for Organizations with the Generation Operationalization and Discontinuation Organization. In: Winter, R., Zhao, J.L., Aier, S. (eds.) DESRIST 2010. LNCS, vol. 6105, pp. 226–241. Springer, Heidelberg (2010)
10. Cameron, B., Leaver, S., Worthington, B.: Value-Based Communication Boosts Business' Perception of IT. Forrester Research (2009)
11. Gordijn, J.: Value-based requirements Engineering: Exploring innovatie e-commerce ideas. Vrije Universiteit Amsterdam, Amsterdam (2002)
12. Aveiro, D.: G.O.D. (Generation, Operationalization & Discontinuation) and Control (sub)organizations: A DEMO-based approach for continuous real-time management of organizational change caused by exceptions. UTL, Lisboa (2010)
13. Pombinho, J., Tribolet, J.: Modeling the Value of a System's Production – Matching DEMO and e3Value. In: 6th International Workshop on Value Modeling and Business Ontology, Vienna, Austria (2012)
14. Pombinho, J., Aveiro, D., Tribolet, J.: Business Service Definition in Enterprise Engineering - A Value-oriented Approach. In: 4th Workshop on Service Oriented Enterprise Architecture for Enterprise Engineering, Beijing, China (2012)

Towards Developing a Model-Based Decision Support Method for Enterprise Restructuring

Eduard Babkin and Alexey Sergeev

National Research University – Higher School of Economics
Dept. of Information Systems and Technologies,
Bol. Pecherskaya 25,
603155 Nizhny Novgorod, Russia
eababkin@hse.ru, aisergeev@yahoo.com

Abstract. In modern world enterprises need to be agile in their operation and structure to react to changes quickly. One of the open questions here is how to develop the enterprise, or, to be more precise, if enterprise needs to be developed, and if yes, in which way. In this research we are focusing on the case when enterprise stakeholders understand the need of enterprise development, have ideas for that, and they need decision support method to understand if enterprise restructuring is likely to be successful and cost effective. Another covered topic is how to choose the best option for restructuring from variety provided. In this paper we describe the developed decision support method which combines DEMO methodology and transaction costs theory for quantitative costs estimation. To make this method applicable and reproducible we proposed few enhancements to DEMO notation.

Keywords: DEMO methodology, transaction cost theory, decision support, enterprise restructuring.

1 Introduction

In modern world enterprises should react to changes quickly, be agile and increase velocity in their actions [4, 5]. Enterprises should evolve to secure their position in the market, to take new opportunities. One of the problems here is how to define which changes are necessary to apply to the structure of the enterprise, to its business functions and operations. To be more precise, often enterprise needs to restructure itself - either to split itself into parts or to merge with another company.

Several works were done [8, 9] which apply principles of Enterprise Engineering [5], Enterprise Ontology and DEMO [2, 3] to rigorous studies of structural changes during enterprise splitting or merging. However there is still a lack of quantitative methods which facilitate comprehensive and objective evaluation of several alternatives possible for reengineering of enterprise structure.

The goal of this research is to develop a reusable method of decision support for defining the best solution for splitting or merging of the enterprise. The idea is to use combination of DEMO methodology and transaction costs theory to build a uniform

H.A. Proper, D. Aveiro, and K. Gaaloul (Eds.): EEWC 2013, LNBIP 146, pp. 17–27, 2013.

quantitative method to be used by enterprise decision makers to choose the best option for splitting or merging from the variety identified. Comparing to other methods, the method proposed in this work is the quantitative one, which means more practical relevance for enterprises. To add to this, it introduces a usage of combination of DEMO methodology and transaction costs theory, which have never been done before.

The research is based on the empirical observations and data available from the real-life case of a local car service company. In that case the company owners need a reliable decision support method for choosing a direction of further business development. As for the options, they see one way of splitting the enterprise and three ways of creating new department within its structure. All this led to two separate cases which needed to be worked out:

- Enterprise structure simplification. It includes spin-off of some departments into separate legal entities to mitigate risks and costs, closing some departments as unprofitable and non-relevant, and closing some departments with vital functions to enter into a contract with another company which will fulfil these functions to decrease costs and increase overall business operations efficiency.
- Enterprise structure complication. It includes merging with another company, and opening new department with new business functions.

To provide a reliable decision tool for stakeholders of the case we apply principles of Enterprise Ontology and DEMO modelling together with developments of transaction costs theory. As a result we came up with the proposal on how to enhance DEMO notation and developed a method of costs estimation for changes of enterprise structure. Our method consists of the set of actions which should be taken in order to estimate the ease and costs of enterprise structure changes. This method is supposed to be used in consulting by experienced enterprise modellers. To add to this, our method can be programmed into software application to automate and simplify the process of enterprise structure modelling and costs estimation.

In this article Section 2 describes the details of the case and the business context. Section 3 shows the theoretical background for our research by briefly describing DEMO methodology and transaction costs theory. Section 4 describes our proposals to extend the notation of DEMO in order to include key principles of transaction costs theory and costs estimation. In Section 5 our method of extended DEMO modelling and decision making is presented. Section 6 contains practical results of application of the proposed methods in the case of the car company. In the conclusion we outline major results and open issues for further research.

2   Business Context

Car service company "TSS-Auto" is a small business enterprise employing 22 people. It operates as a fully separate legal entity not dependant from any bigger company. Services provided to customers include car repair, regular car maintenance (obligatory by Russian law), colour matching and colouration, selling of car lubricants.

Owners of a company requested help in choosing a way on how to develop their enterprise any further. The reason for this lies in the situation when the core business of a company could not be grown any further due to several constraints like lack of experienced workforce, tough competition, and some others. So the owners decided to optimize their costs and/or find the ways to enter into supplemental / accompanying business. In other words, they needed a reusable method of decision support for defining best solution for splitting or merging of the enterprise. They came with a set of different thoughts about possible changes in the enterprise.

A meeting with enterprise owners was set in order to list all the possibilities for company change, rank them and choose few most realizable from owners' point of view. The list included:

- Excluding car wash from the functions of the company. While this is the vital function of the company (because one cannot start repairing or colouring the car before it washed), owners are thinking about excluding this due to high costs. The root cause for this lies in the fact that there are no water plumbing exists to the company building. It means that company has to buy water separately and deliver it to the premises, so it causes the high costs and poor quality of car wash. The idea is to find a partner which locates close to the company building and will be able to wash cars regularly. This case obviously means enterprise structure simplification by closing one of the functions.
- Taking functions of car insurance agent. Car service is dealing with many customers' cars and has connections to insurance companies (servicing its customers), so adding function of car insurance agent should be easy. But company owners want to understand how it will map into current enterprise structure and what the estimated costs for this action are. Choosing this way of developments means enterprise structure complication by adding new function / department.
- Starting to sell auto parts directly to customers (which may mean becoming an official / certified dealer for some of the auto parts manufacturer). Currently company sells parts to customers indirectly – only then they are included into repairing process. The idea is to start selling parts directly to customers, when they can walk into some kind of shop and buy parts without requirement to repair car in this car service. This change may be very serious actually, because it requires creating new business relations to parts manufacturers, hiring new people, etc. The request was to estimate costs and ease of integration of this change into enterprise structure. Therefore, it means enterprise structure complication.
- Starting to sell used cars. Whilst this business is close to car service for some extent (some cars are 'fully' crashed, so many car owners are glad to sell it for parts or for repairing and future reselling; both can be easier and more effectively done by car service itself), it is also risky and money-consuming. That is why company owners wanted to be as much informed as possible about consequences of this change. It lies into the enterprise structure complication case.

Although the stakeholders do not consider directly an opportunity of merging with another company such variant of actions should be included to the set of feasible decisions. We will cover this topic in the theory part, and our proposed method should be relevant for this case as well, but we cannot check it due to requirements from real business case.

3   Theory Basis

Our research is based on two principal foundations, namely DEMO Enterprise modelling and transaction costs theory (TCT).

DEMO (Design & Engineering Methodology for Organizations) is a methodology for the design, engineering, and implementation of organizations and networks of organizations. The entering into and complying with commitments is the operational principle for each organization. These commitments are established in the communication between social individuals, i.e. human beings [1].

In DEMO the basic pattern of a business transaction is composed of the following three phases. An actagenic phase during which a client requests a fact from the supplier agent. The action execution which will generate the required fact. A factagenic phase, which leads the client to accept the results reported.

Basic transactions can be composed to account for complex transactions. The DEMO methodology gives the analyst an understanding of the business processes of the organization, as well as the agents involved. The analysis of models built on the methodology of DEMO allows the company to obtain detailed understanding of the processes of governance and cooperation and serves as a basis for business reengineering and information infrastructure development, consistent with the business requirements.

DEMO models are very compact and take no more than a sheet of paper. Analyst does not need any special tools, in some cases a pen and paper is enough.

Each DEMO model has practical importance and useful for the analysis of the whole organization. The analyst can build all five DEMO models for each business process, but the study found out that the process model, the action model and the state model do not bring appreciable benefits in terms of our research.

The interaction model (IAM) is the most compact ontological model. It shows area of responsibility of actors and the transactions that are important in terms of business operation. That is why we choose IAM for the modelling of the case in our research. All the cases of DEMO usage are shown in the next sections of this paper.

One of the features of our decision support method described further is the usage of transaction costs theory (TCT) for enterprise restructuring and operation costs estimation. In economics and related disciplines, a transaction cost is a cost incurred in making an economic exchange (in other words, it is the cost of participating in a market). Usually transaction costs are divided into three rather broad categories [6, 7]:

1. Search and information costs. As included in the name of this type of costs, they are related to searching, obtaining and using information. For example, price or market research can be included here, as well as costs incurred in

determining if some good has required quality, available for order, has the acceptable price, etc.

2. Bargaining costs. These are the costs required to come to an agreement with the other party (parties) to the transaction, creating an appropriate contract form, etc. It may include bills for business meals with partners, payments to lawyers for contract form preparations, and so on.

3. Policing and enforcement costs. These are the costs of making sure the other party is following the terms of the contract, and in case of any violation it includes taking appropriate actions. This type of costs includes monitoring of work processes (if they are in accordance to the contract), all fees associated with lawsuits, etc.

Literature related to transaction costs theory highlighting many examples of transaction costs types which may occur during enterprise operation. Also needs to be said that some types of transactions costs can be typical for selected country, business segment of enterprise operation, period of economic cycle, etc. However, there is still an open question how to estimate transaction costs correctly. Further in this paper we propose ways on how we can guarantee our transaction costs estimation.

4   Integration of DEMO and TCT in the Proposed Method

To be effective and accepted by company stakeholders the developed decision support method should radically simplify estimation of costs for enterprise restructuring and future operations in case new functions were included. Overall, we assume that we need to estimate cost of restructuring itself (using theory of transaction costs), and then costs of operation of new enterprise structure (to understand if it is better than initial one in the long term).

From that point of view DEMO is good for separating transactions and actor roles for execution of each transaction. Therefore it provides good background for separation of costs of each transaction and using transaction costs theory at the same time.

The labour cost is a big part of cost-of-production, so as the first step of developing our new method of decision support we propose to estimate changes within actor role and function role enterprise structure using the following table which we call "Actor-Function Role Table":

Table 1. Actor-Function Role Table

	Number of employees	Average salary	Actor_role_1	Actor_role_2	Actor_role_3
Function_role_1	2	23000	30%		70%
Function_role_2	3	36000		100%	
Function_role_3	6	18500		10%	90%
Total cost			2*23000*0.3=13800	119100	132100

Under "Function role" we understand employees' roles as mentioned in their job contract or as understood under their job responsibilities. Under "Actor role" we understand actor roles as per DEMO methodology. As DEMO focuses on construction of the enterprise, we need to elaborate more on the link between function roles and actor roles as we understand it. Obviously, each employee has its function role as described in its job contract or as understood by job responsibilities. At the same time, each employee is fulfilling certain actor roles (either ontological, infological or datalogical), and, as based on our experience, employees with the same function roles are fulfilling the same actor roles. To add to this, each function role usually matches few different actor roles. Based on this, we can find out the percentage of work time each function role spends fulfilling certain actor role. It helps to estimate costs in terms of employees' salaries associated with each actor role.

Number of employees shows the number of people having certain function role, whose average salary is mentioned in the next column.

Using this table we are able to calculate costs of each actor role in the organization (for a period of a month, for example). Creating this table for the enterprise structure before and after change is applied; we can compare the labour costs.

As the next step, we are able to estimate cost of each transaction (either ontological, infological or datalogical) combining cost-of-production and transaction costs theories. As a result of the exploration, we came with the following table to be filled in (we call it "Transaction Cost Table"):

Table 2. Transaction Cost Table

Cost name	Cost estimation
Cost of production (materials, electricity, etc)	Can be calculated easily
Different kinds of transaction costs, applicable for current situation	Probability/risk estimation in money equivalent

Similar to that, using the theory of transaction costs, we can estimate costs of changing enterprise structure (we call this "Restructuring Cost Table"):

Table 3. Restructuring Cost Table

Cost name	Cost estimation
Hiring new people	
Firing people	
Payment to lawyer for creating contract	
… + different kinds of transaction costs, applicable for current situation	

During our research we observed two typical patterns in enterprise restructuring.

The first one is that after restructuring affecting many transactions and actor roles (many transactions and actor roles are eliminated or otherwise – added) usually one new transaction appears – "keeping enterprise operation in accordance to DEMO model". It means that after restructuring time and efforts need to be invested in constant monitoring of how enterprise operates in relation to created DEMO model, and some corrective actions sometimes need to be taken in order to correct actors' behaviour and keep everything as planned.

The second pattern is the fact that when transaction is moved from inside enterprise to enterprise border (which means that now collaboration between external and internal actors is needed), many new risks, and therefore transaction costs immediately appear, like risks of lawsuits (and as a result payments to lawyers as a costs), risk of losing a partner (and therefore costs of searching for new partners), etc.

As a guarantee of the correctness of costs estimation in both types of tables above, we can propose the following methods:

1. Costs are first estimated by 2 groups of experts separately, and then costs estimations are compared. For the cases of contradictions both groups of experts are working together in order to reach a consensus.
2. Using statistics data for the area of business operation of the enterprise in order to get costs estimations using theory of mathematical statistics. The main drawback of this approach is that not all the required data may exist or to be publicly open for usage.
3. Using imitating modelling techniques for creating and running enterprise model to estimate its behaviour and possible costs.

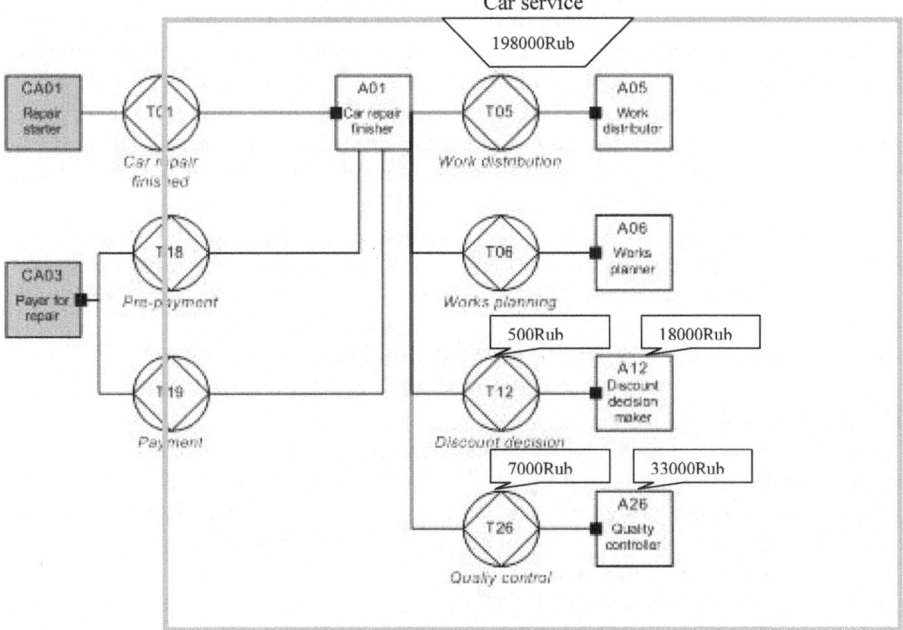

Fig. 1. Example of enhanced ATD

Methods proposed above can be used altogether in any combinations. In current research we used combination of the methods 1 and 2.

After fulfilling all the tables above, we can map this info to Actor Transaction Diagram or Organization Construction Diagram. To each actor role on the diagram we put estimated costs of operation (per month/year), to each transaction we add estimated costs of operation as well. To the organization border we add costs of enterprise restructuring. Example (fragment of ATD):

This enhanced diagram explicitly shows costs associated with the changes to the enterprise structure. Comparing diagrams for each change can be done much easier and for less time.

5   Modelling Method

Based on our research, we propose a reproducible method for choosing the best way of developing the enterprise – either simplification or complication of its structure.

Step 1. Company owners, management or analysts propose options on how it is possible to change the enterprise from their point of view. This call should be made from business side as the subject matter experts. As a result of this step we expect to get the list of possible changes to the enterprise. Based on our experience, brainstorming is the best technique to be used by company representatives to come up with the ideas for changes. As a criterion for success we propose having the list of possible changes to enterprise structure agreed by all stakeholders

Step 2. Meeting with stakeholders and subject matter experts to rank all possibilities and choose only few of them for further development. People invited to this meeting may be different from those who were proposing changes on the Step 1. The main goal of the meeting is to discuss ease of implementation, opportunities and outcomes associated with each possible change to focus modelling efforts only on those changes which are the most relevant according to enterprise representatives. As a result of this meeting we expect to define very limited list of structure changes to be modelled. As a criterion of success we understand the short list of possible changes agreed by all company stakeholders and business modellers.

Step 3. Ontological (and, potentially, infological and datalogical – depending on the level of company requirements) model of the current enterprise structure is created using DEMO methodology. As a result of this step we have Actor Transaction Diagram (or Organization Construction Diagram) created for current enterprise structure. This diagram needs to be verified with enterprise stakeholders. A success criterion for this step is the diagram (either ATD or OCD) fully created and agreed with enterprise stakeholders.

Step 4. The same models of enterprise are created for each potential structure change. It means that on this step we need to create the same diagrams (either ATD or OCD) for each potential enterprise structure after changes applied. Again, it needs to be verified from business perspective with enterprise stakeholders. A success criterion is the same as on the previous step.

Step 5. Applying extensions of DEMO model proposed in Section 3. On this step we are supposed to estimate costs of restructuring of the enterprise and costs of transactions after changes applied. As described in Section 3, the idea is to create comprehensive lists of possible costs including developments of transaction costs theory. After such lists are created for both restructuring itself and for each transaction which was affected by the change, next step will be to estimate all costs that possible. Obviously, some costs can be estimated only by subject matter experts, but the main goal of this step is to simplify costs estimation for enterprise stakeholders as much as possible. A success criterion for this step is the completeness of each list of costs, which means that no or minimum of cost types will be added to the lists on the Step 6. Also, as a criterion of success for this step we propose estimation of as many costs as possible before presenting to company stakeholders.

Step 6. Presenting results to company stakeholders and subject matter experts so they can finalise costs estimation. On this step enterprise stakeholders and subject matter experts are supposed to fill the gaps in the costs estimation to get the final picture of costs associated with each change. As a result we get the final enhanced ATD or OCD diagrams (as per Section 3) for each change. As a success criterion for this step we propose having these diagrams fully ready for comparison on next step.

Step 7. Comparing initial enterprise structure and costs, and each enterprise structure after change to choose the best solution based on ease and of implementation and cost of future operation. Using enhanced ATD or OCD, we can compare initial enterprise structure and each structure after change to understand all costs associated with changes. It should lead to understanding which enterprise restructuring will be affordable in current circumstances, will lead to reasonable addition to costs of operation comparing to profits planned by enterprise stakeholders. As a result of this step, the best way of changes and course of actions should be defined together by company stakeholders and enterprise modellers to be implemented on the next step. As a success criterion we propose the ease and comprehensiveness of comparison of each case of changes.

Step 8. Implementation of the solution chosen under control of enterprise modeller to keep all the changes as per planned enterprise structure. The participation of enterprise modeller is needed to keep all changes within planned enterprise structure to keep costs close to estimations. As a result of this step changes should be implemented. A success criterion is to have final enterprise structure the same as planned, and final costs as close to estimation as possible.

6   Application

In our case we achieved several practical results. Following our method, on Step 1 company representatives provided 9 ways of changing enterprise structure. Each possible change was discussed and agreed by company stakeholders internally, and then presented to researchers. On Step 2, during the meeting with enterprise stakeholders, we chose 4 changes to be modelled (as described in Section 2). All four possible ways of changes were agreed between company stakeholders and researchers. On step 3 we created ontological model of the enterprise expressed in Actor Transaction Diagram, which was verified by company owners. Enterprise stakeholders were not requesting creation of infological and datalogical models in this case. On step 4 we created four ATD diagrams for each potential structure change. During this step we noticed that in case of enterprise structure complication it is easier to imagine that we are merging two enterprises. For example, in case of adding functions of car insurance agent, it was easier to consider car insurance agent as a separate enterprise first, create its ontological model expressed in ATD, and then merge its ATD with initial ATD diagram of our enterprise adding necessary changes. After that ATD diagrams were verified with company representatives. On Step 5 we applied proposed enhancements to DEMO notation. Taking into account rather small number of employees in the company under consideration, it was easy to fill Actor-Function Role Table. Next we were able to fill Transaction Cost Table and Restructuring Cost Table, except estimating few types of costs which required knowledge of subject matter experts.

As the next step we are planning a meeting with company stakeholders to present the results of our modelling and costs estimation efforts.

During our work on this case we noticed that many actions can be simplified and automated by software tool. As such, this tool can include templates for proposed enhancements to DEMO notation and the comprehensive list of transaction costs as proposed in literature (so modeller can just choose applicable types of costs).

7   Conclusion

A new method of decision support during re-engineering of enterprise structure was presented. This method combines DEMO and transactions costs theory for objective quantitative evaluation of several alternatives during splitting or merging small or medium enterprises in a complex business context. Applicability of the method proposed was demonstrated in the case of real car company which faces business challenges and needs re-engineering.

The proposed method has several positive features such as using DEMO methodology for enterprise structure modelling, which consumes around 90% less time comparing to other methods [1,2]. To add to this, our method is easily reproducible and can be applied regardless business segment of the enterprise.

In comparison with other approaches (such as [3]) our original method uses quantitative metrics to estimate enterprise restructuring and future operation costs, which leads to better understanding by enterprise stakeholders and more accurate planning of changes.

During development of the method the original notation of DEMO was modified in order to include developed Actor-Function Role Table, Transaction Costs Table and Restructuring Cost Table and was enhanced by modifying Actor Transaction Diagram (or Organization Construction Diagram). It helps to graphically represent all the changes and costs associated with enterprise restructuring.

As a way for future development of proposed method we consider developing software tool which will help to automate and simplify the method application. It will not also reduce enterprise modeller's time for creating DEMO diagrams and estimations costs, but will also give an opportunity to enterprise stakeholders to use this tool in future by their own and not to involve 3rd-party enterprise modeller.

This work was supported by National Program of Research and Development (State Contract # 14.514.11.4065).

References

1. Dietz, J.L.G.: Enterprise Ontology: Theory and Methodology. Springer, Heidelberg (2006) ISBN-10 3-540-29169-5
2. Op 't Land, M., Dietz, J.L.G.: Benefits of Enterprise Ontology in Governing Complex Enterprise Transformations. In: Albani, A., Aveiro, D., Barjis, J. (eds.) EEWC 2012. LNBIP, vol. 110, pp. 77–92. Springer, Heidelberg (2012)
3. Op 't Land, M.: Towards Evidence Based Splitting of Organizations. In: Ralyté, J., Brinkkemper, S., Henderson-Sellers, B. (eds.) Situational Method Engineering: Fundamentals and Experiences. IFIP, vol. 244, pp. 328–342. Springer, Boston (2007)
4. Op 't Land, M.: Applying Architecture and Ontology to the Splitting and Allying of Enterprises: Problem Definition and Research Approach. In: Meersman, R., Tari, Z., Herrero, P. (eds.) OTM Workshops 2006. LNCS, vol. 4278, pp. 1419–1428. Springer, Heidelberg (2006)
5. Op 't Land, M., Proper, E., Waage, M., Cloo, J., Steghuis, C.: Enterprise Architecture, ch. 2. Springer (2009)
6. Commons, J.R.: Institutional Economics. American Economic Review 21, 648–657 (1931)
7. Cheung, S.N.S.: Economic organization and transaction costs. In: The New Palgrave: A Dictionary of Economics, vol. 2, pp. 55–58 (1987)
8. Harmsen, F., Proper, H.A.E., Kok, N.: Informed governance of enterprise transformations. In: Proper, E., Harmsen, F., Dietz, J.L.G. (eds.) PRET 2009. LNBIP, vol. 28, pp. 155–180. Springer, Heidelberg (2009)
9. Proper, H.A., Op 't Land, M.: Lines in the water: The line of reasoning in an enterprise engineering case study from the public sector. In: Harmsen, F., Proper, E., Schalkwijk, F., Barjis, J., Overbeek, S. (eds.) PRET 2010. LNBIP, vol. 69, pp. 193–216. Springer, Heidelberg (2010)

Exploring Organizational Implementation Fundamentals

Martin Op 't Land[1,2,3] and Marien Krouwel[1,4]

[1] Capgemini Netherlands, P.O. Box 2575, 3500 GN Utrecht, The Netherlands
{Martin.OptLand,Marien.Krouwel}@capgemini.com
[2] Antwerp Management School, Sint-Jacobsmarkt 9-13, 2000 Antwerp, Belgium
[3] Delft University of Technology, P.O. Box 5031, 2600 GA Delft, The Netherlands
[4] University of Antwerp, Prinsstraat 13, 2000 Antwerp, Belgium

Abstract. To survive and even thrive on environmental and internal change, organizations have to be agile. Though change occurs in organizational essence, such as products and services delivered, most of the time change deals with different organizational implementations, such as sourcing, order of working and distribution of tasks. To informedly decide upon such organizational implementations, a systematic overview of organization implementation variables is required, which is currently not available. We drafted a list of organization implementation variables from literature, and tested it against two different organization implementation descriptions of OMG's EU-Rent case, using the DEMO model for this fictitious car rental company as its implementation independent essence. We found a list of 20 of such variables from literature, which was extended in the two tests by another 10 variables. Using these variables in Enterprise Engineering enables traceability in governing enterprise transformations; moreover, we expect many of them to have the potential to be generically supported by IT, thus enabling agile IT.

Keywords: DEMO, Agile Enterprise Engineering, Enterprise Ontology.

1 Introduction

As strategic and operating conditions become increasingly turbulent due to factors such as hyper-competition, increasing demands from customers, regulatory changes, and technological advancements, the ability to change becomes an important determinant of firm success [1]. This ability is generally referred to as *agility*, e.g., as summarized by Oosterhout [2]: "Business agility is the ability of an organization to swiftly change businesses and business processes beyond the normal level of flexibility to effectively manage highly uncertain and unexpected but potentially consequential internal and external events, based on the capabilities to sense, respond and learn."

Though change – as a consequence of external and internal events – occurs in organizational essence, such as products and services delivered, most of the time change deals with different implementations [3]. Typical organizational implementation choices include sourcing, order of working and distribution of tasks.

H.A. Proper, D. Aveiro, and K. Gaaloul (Eds.): EEWC 2013, LNBIP 146, pp. 28–42, 2013.

To informedly decide upon such organizational implementations, a systematic overview of organization implementation variables is required, which is currently not available. Research in the agility domain appears to focus on black-box variables – such as the type of events triggering change [4] or the measurement of agility [5] – or on the transformation processes needed to bring about the change [6]. Research in the agility domain about white-box variables is until now restricted to the IT domain; e.g., Normalized Systems theory [7] proposes a "Set of Anticipated Changes" for IT systems in terms of the (detailed) function of an IT system, such as an additional data field or an additional trigger element.

We drafted a list of organization implementation variables from literature, and tested it against two different organization implementation descriptions of OMG's EU-Rent case [8]. For this fictitious car rental company, we used its DEMO model [9] as its implementation independent essence, since it remains the same as long as the products and services of an enterprise stay the same.

We found a list of 20 of such variables from literature, which was extended in the two tests by another 10 variables. Examples of these variables include (a) the choice to combine or split actor roles in the work of an employee, (b) to apply delegation and separation of functions, and (c) to apply a fixed order of working in a process or allow steering of that order by the individual employee.

Explicitly using these variables in Enterprise Engineering enables traceability in governing enterprise transformations. Also we expect many of them to have the potential to be generically supported by IT, thus enabling agile IT. For instance, when the choice how to combine or split actor roles in the work of an employee (ad *a*) can be registered explicitly and in one place, all connected software applications can use this information to change their – e.g., GUI and security – behavior accordingly, potentially without the need for this software to be reprogrammed when this choice is changing. Since changing (also organizational) implementation variables tend to have combinatorial effects [7], the future potential for wider validation and application of this list is significant.

The remainder of this paper is structured as follows. Section 2 elaborates the problem statement: what do we understand by implementation and agility, and why is it relevant to find organization implementation variables. Section 3 describes the draft list of organization implementation variables from literature, which is then tested in Section 4, using the two different organizational implementations of the EU-Rent case. Finally, Section 5 provides the conclusions as well as directions for further research.

2 Problem Statement

In this section we will first introduce some definitions by explaining the Generic System Development Process (GSDP) as defined by Dietz (Fig. 1). After that, we will present our findings on existing literature about agility, events and implementation.

2.1   Enterprises and Generic System Development Process

We define *enterprise* as a goal-oriented cooperative. The organization of an enterprise is a heterogeneous system, constituted as the layered integration of three aspect systems, namely the Business (B) system, the Informational (I) system and the Documental (D) system [10, p115]. The production of these systems concern (B) original acts (material and immaterial), such as deciding, judging and creating, (I) informational acts, such as remembering, recalling and computing and (D) documental acts, such as storing, retrieving, transmitting and copying.

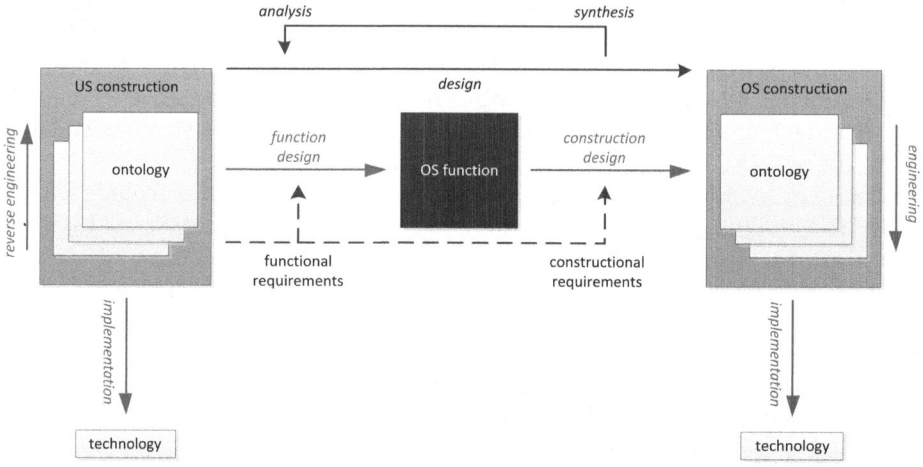

Fig. 1. Generic System Development Process [11]

Any development process concerns two systems involved, the *Object System (OS)*, and the *Using System (US)*. The OS is the system to be developed; the US is the system that will use the services (the functionality) offered by the OS once it is operational. The development of the OS consists of a *design*, an *engineering*, and an *implementation* phase. The design phase comprises a function design and construction design. *Function design*, the first phase in the design of the OS, starts from the construction of the US and ends with the function of the OS. Function design delivers the requirements of the OS, or a black-box model of the OS. This black-box model clarifies the behavior of the OS in terms of (functional) relationships between input and output of the OS. This function model of the OS does not contain any information about the construction of the OS. *Construction design*, the second phase in the design of the OS, starts with the specified function of the OS and ends with the construction of the OS. Construction design bridges the mental gap between function and construction, which means establishing a correspondence between systems of different categories: the category of the US (where the function of the OS is defined), and the category of the OS. Construction design delivers an ontology, the highest level white-box model of the OS. This white-box model clarifies the internal construction and operation of the system in terms of collaboration between its elements to deliver products to its environment. By an *ontology* or ontological model of a system we understand a model of its construction that is completely independent of the way in which it is realized and implemented. The *engineering*[1] of a system is the process in which a number of white-box models are produced, such that every model is fully derivable from the previous one and the available specifications. Engineering starts from the ontological model, produces a set of subsequently more detailed white-box models and ends with the implementation model. By *implementation* is understood the assignment of technological means to the elements in the implementation model, so that the system can be put into operation. By *technology* we understand the technological means by which a system is implemented. A wide range of technological means is available, including human beings and organizational entities, ICT artifacts (e.g., phone, email, computer programs), and mechanical means. By *implementation variables* we mean the dimensions in which organizational implementation choices are made.

As an enterprise consists of three integrated layers, it can be developed by applying the GSDP three times [12]:

1. first the US is the (many times: commercial) environment in which the enterprise is going to be operational, and the OS is the B-organization of which the functional model contains the services that the enterprise will deliver to its customers [10, p77];
2. then the US is the B-organization, and the OS is the I-organization of which the functional model contains the information services (e.g. reason, compute, remember, reproduce) that the I-organization will deliver to the B-organization [10, p114];
3. finally the US is the I-organization, and the OS is the D-organization of which the functional model contains the documental services (e.g. store, retrieve, copy, destroy, transmit) that the D-organization will deliver to the I-organization [10, p114].

By applying GSDP for the enterprise as a whole (so three times), it is now possible to systematically categorize impact of change. Change in environment can be responded to by choices in function, which in turn will influence construction on both ontological and implementation level. Similarly, changes in the B-organization generally will influence the I- and D-organization, and the other way around. We illustrate this by some examples of changes in a law:

- a law stating that one organization cannot provide both banking and insurance services, affects the B-functional model;
- a law stating rules for granting a subsidy affects the business rules, i.e., the B-ontological model;
- a law stating reporting obligations affects the way of providing information and/or saving of data, i.e., the I-functional and D-functional model;
- a law stating the channels offered affects at least the implementation.

---

[1] Engineering is meant here in the narrow sense of the term, contrary to its general use in civic engineering, electrical engineering, mechanical engineering, etc.

Although one is obliged to adhere to legislation, law often leaves freedom of choice. For example, if law states one must at least provide a non-digital channel, one is still free to choose between telephone and physical service desk (or both). So, the change in law is an event in the environment, possibly but not necessarily responded to by an organization with a change in the function and/or construction of the organization.

2.2   Agility, Events and Implementation

To thrive in an environment of continuous and often unanticipated change, an enterprise needs to be agile [5]. Oosterhout [2] summarizes several definitions of agility as "the ability of an organization to swiftly change businesses and business processes beyond the normal level of flexibility to effectively manage highly uncertain and unexpected but potentially consequential internal and external events, based on the capabilities to sense, respond and learn." The question then arises what these kinds of events are.

Using the perspectives of the Enterprise Engineering Framework (EEF) [13], we categorized several event classifications found in literature (Table 1). As literature does not explicitly mention whether Technology and Resources deal with an *available* or a *chosen* implementation, we split EEF's original Context perspective in Context (environment & demand) and Context (supply), and positioned Technology/Resources (available) in the Context (supply) perspective. Likewise, we made a distinction between Customer needs, which is in the Context (demand perspective, and the choice of an enterprise to answer these needs with certain Products and services (supply), residing in the Function perspective. So in the event classes from literature

- 8 concern changes in the context of the organization, and can be reason for change in any aspect of the organization;
- 2 concern changes in the function of the organization, representing the choices in response to the context;
- 3 concern changes in the ontology of the organization, and
- 5 concern the implementation of the organization – 3 for parties and people and 2 for ICT.

Remarkably, no events specific for the informational or documental organization are discerned.

Common definitions of agility emphasize the contextual and functional perspective. Sarkis [5] focuses on *metrics* for agility – such as acquisition time, demand change cost and amount of capable workers on certain equipment – just as Tsourveloudis et al. [15], which propose a set of quantitative agility parameters for calculating the overall agility of an enterprise. Van Oosterhout [2, p216] asks for more research to analyze different types of business agility needs, also because he expects that building IT platforms which support all these types of business agility needs will be very expensive. So, all definitions of agility found (including [16], [4], [6]) are mostly black-box or functional, i.e., they agree that

Table 1. Categorization of event classifications in EEF's perspectives (adapted)

Context (environment & demand)		Catastrophic [2]
		Social/legal [2, 4, 5]
		Business network [2]
		Competition [2, 4]
		Customer needs [2, 4, 5]
Function		Products and services (supply) [5]
		Quality of Service (QoS) [14]
Construction — Ontology		Processes (business rules) [5]
		Technology (methods) [4]
		Internal change [2, 4]
Construction — Implementation	Parties and People	Resources [5]
		Processes (responsibility) [5]
		Internal change [2, 4]
	ICT and other means	Technology [2, 4, 5]
		Internal change [2, 4]
Context (supply)	Parties and People	(available) Resources [5]
		Social (workforce expectations) [4]
	ICT and other means	(available) Technology [2, 4, 5]

one should be able to change; they do not tell, white-box or constructional, in what respect an organization should be able to change.

Directed searches for organization implementation (variables) did not yield anything useful. On top of earlier mentioned literature, Google Scholar searches (in English and Dutch) were performed with the terms organization(al) implementation, organization(al) change, organization(al) aspects, organization(al) dimensions, organization(al) design, and (organization(al)) implementation aspects.

The only frameworks that seem to deal with organization implementation, using other terminology however, are the COPAFILTH[2] framework [17, 19] and Hoogervorsts Framework for Enterprise Engineering [18], summarized in Table 2. Half of the aspects mentioned still concern the environment or functional perspective of the organization, the other half concerns the level of construction, of which 3 categories deal with organizational implementation. In each of these categories some examples are mentioned, however no exhaustive list is provided.

So, our broad research question is to have a complete list of organization implementation variables. In this paper we will create a first set of such variables.

2.3 Approach

We will first categorize the variables found in literature in the system types of EEF [13]. Secondly, we will test (deductive step) the completeness of these variables against two different organization implementation descriptions of OMG's

---

[2] Translation of Dutch COPAFIJTH.

Table 2. Categorization of COPAFILTH [17] and Framework for Enterprise Engineering [18] in EEF (adapted)

			Business	Informational	Documental
Context (environment & demand)			Commerce/Business (demand) [17, 18]		
			Legal [17]		
Function			Commerce/Business (products and services) [17, 18]		
			Organization (flexibility) [17, 18]		
			Administrative (management) [17]		
			Finance [17]		
			Technology (quality and flexibility [17], security [18])		
				Information (supply, quality) [17, 18]	
Construction / Implementation	Ontology		Administrative (structure) [17]		
			Information (need) [17, 18]	Information (structure) [18]	
		Parties and People	Organization (structure and culture [17], culture and processes and employees [18])		
			Personnel [17]		
			Administrative (order) [17]		
		ICT and other means	Technology [17, 18]		
			Organization (technology, means) [18]		
			Housing [17]		
					Information (gathering, storage, distribution) [18]
Context (supply)	Parties and People		Business (suppliers, partners) [18]		
	ICT and other means				

EU-Rent case [8], and possibly extend the list (inductive step). In this paper we will not elaborate the motives to give these variables a certain value – e.g., choose for a specific organizational split [20] – or the coherence of these variables – e.g., more ICT could influence the amount of personnel; we stop at the level of identifying the variables.

3 Variables from Literature

Table 3 summarizes the organization implementation variables as found in our literature search (see subsection 2.2). Recent research in Adaptive Case Management (ACM) [21, 23] elicits the following nuances in these variables.

- some business rules are optional, others are mandatory; it should be possible to document the (lack of) complying to these rules in an *execution trace*;

Table 3. Organization implementation variables in EEF (adapted)

		Business	Informational	Documental
Construction / Ontology		Actor roles, transaction kinds, information links [10,17,18][a]		
			Business rules [5,10], methods [4]	
Construction / Implementation / Parties and People			Organization structure [17]	
			Departments	
			Functionary types[b] [17]	
			Delegation [5]	
			Separation of function[c] [17]	
			Order of working [17,18]	
			Assignment of tasks[d] [5,17]	
			#Full-time Equivalent (FTE) [5]	
			Skills and competences [5,17]	
			Sourcing [2,5]	
				Language support
				Data structure[e] [18]
Construction / Implementation / ICT and other means			Locations of offices [17]	
			Equipment and infrastructure[f] [2,4,5,17,18]	
			WFMS and execution trace [21]	
			Degree of automation [17]	
			IT Integration level [5,17]	
				External data sources [22]
				D(B)MS, CMS
				Channels[g] [17]

[a] In the Informational world these concern information needed for B-actors, management information, term monitoring, reporting, etc.

[b] including mapping of actor roles (responsibility) to functionary types

[c] e.g. splitting of different steps over different functionary types/persons, or 4-eye principle

[d] top-down, self-regulatory, priorities, teams or individuals

[e] non-structured – e.g., tape (audio/video), (Word, Excel) documents – or structured – e.g., database, XML

[f] Including Man-Machine Interface and GUI

[g] web/email, phone/sms, paper, . . .

– some organization implementation variables get their value at the very last moment, even when the process is running; e.g.,

- which process steps should be performed next in the dealing with this specific case – the so-called *dynamic working plan* as opposed to a fixed work flow,
- who (which person, team, department or even external organization) is going to perform a certain process step, and
- what source of data is sufficient to perform a certain task – e.g., a salary-statement or a bank statement to establish credibility.

4   Validation

In this section we will present the organization implementation variable analysis of two different organization implementation descriptions of OMG's EU-Rent case [8] which will then be compared to the variables found earlier (Table 3). The ontology of the B-organization of this case is presented in Fig. 2. A reading guide for this model can be found in [9].

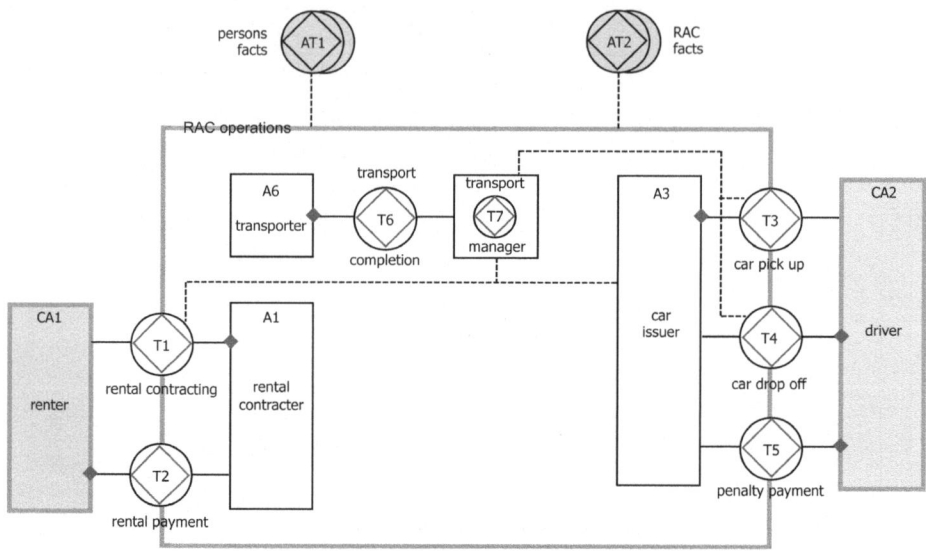

transaction kind	product kind
T1 rental contracting	P1 Rental is contracted
T2 rental payment	P2 the rent of Rental is paid
T3 car pick up	P3 the car of Rental is picked up
T4 car drop off	P4 the car of Rental is dropped off
T5 penalty payment	P5 the penalty of Rental is paid
T6 transport completion	P6 Transport is completed
T7 transport management	P7 transport management for Day is done

Fig. 2. OCD and TPT of Rent-A-Car (adapted from [24])

The first analysis was performed on the Rent-A-Car description [24] (descr. 1). Below, we will show per sentence the organization implementation variable(s) found. For the length of this paper, we cannot present the complete description but summarized the findings in Table 4. The results of the second analysis, performed on the Mini EU-Rent Business Model [25] (descr. 2), can be found in the same table.

Rent-A-Car (or RAC for short) is a company that rents cars to persons, both private ones and representatives of legal bodies, like companies.

This line states that "Rent-A-Car is a company" but it could also have been a network of companies, carrying the same name and embodied in different legal entities. We conclude as implementation variable the embodiment of the ontological model by organizational and legal entities (V1).

It was founded by the twin brothers Janno and Ties back in the eighties.

Contextual sentence, no implementation variables.

They started to hire out their own (two) cars, and they were among the first companies that allowed cars to be dropped-off in a different location than they were picked-up.

From this we find the need for a transportation function between locations which will influence the ontological model of the B-organization. However, no implementation variables are found in this sentence.

To this end, Janno and Ties had made agreements with students in several cities.

We read students are hired to perform some task. So we have an implementation variable regarding the employees in the organization (V2).

For a small amount of money, a student would await the arrival of a rented car, e.g. at an airport, and drive it back to the office of RAC, after which the student would go home by public transport.

From this line we read several things. First, the drop-off location could be anywhere (airport departure hall 3, town center, ...) and not necessarily a RAC office. This implies that the state and accept of the drop-off can happen at any location. For that, the locations of performing certain acts must be defined. Secondly, students are authorized to accept the drop-off, so there is an assignment between employees and act types (during some time frame), and, as the student is not the requester of the drop-off, there is some form of delegation. This implies the students need the relevant information to be available on location, need the right competences to perform this type of task, and possibly need facilities to record the data created. In summary, we found the following implementation variables.

V3: Denotation (syntax) and accuracy of entity types;
V4: Workplace;
V5: Cross-reference which act type can be performed on which location;
V6: Cross-reference which employee is allowed to perform which type of act;
V7: Delegation of act types from functionary type to other functionary type;
V8: Competences/certification.

From the variables regarding employees, location and assignment of tasks (including delegation), the need for information and data recording per employee and per location can be determined. When employees perform different type of acts (possibly involving different transaction kinds), one possibly wants to combine the I- and D-functions to support these acts in one information product.

For example, a student is allowed to perform both drop-offs (accept), pick-ups (state) and transports. If he gets his tasks for a day presented in three lists, he will have to sort out in which order he has to perform his tasks. If he gets this information presented in some Google Maps overview with time tables, in which he can also confirm the end of performing a task, he will better be able to plan the order of performing his tasks.

> *Currently, RAC operates from over fifty geographically dispersed branches in Europe.*

Obviously, again there is the implementation variable about workplace and which type of act is performed at which location – e.g., pick-up can only be done at branches near airports, while drop-off can be done at any branch. However, several questions then arise:

- Who is the addressee of a coordination act (C-act, e.g., request, promise) that is directed to RAC, e.g. the request for rental start? Is it the legal entity RAC, is it a specific branch, or is it an employee at some branch?
- Can a customer request a contract at branch A while the pick-up is done at branch B? Can branch A promise a car rental while the pick-up is done at branch B?
- And if it is necessary for branches to share data, will data be stored locally, centralized or in the cloud?
- Will offices be supported by IT locally (with possibly different systems), shared (using the same systems but locally), or centralized?

From this we summarized the following implementation variables:

V9: Specificity of C-act addressee;
V10: The extent to which the execution of acts within one event, is restricted to the location at which the event is triggered;
V11: Location of data storage (local, centralized, cloud);
V12: Applications, including at which locations.

> *Many cities have a branch, some even several, and there are branches located near all airports. One of the branches is the original office where Janno and Ties started and where both are still around. Being mechanical engineer by education, they have kept loving to drive and maintain cars, even since they are the managing directors of a million euro company.*

Context, no implementation variables.

> *The head of the front office of the home branch is Chiara.*

First, we recognize the notion of departments (front office) and organizational hierarchy. Secondly, we see the notion of functionary type (head of front office), the fulfillment of functionary types by employees (Chiara is head of front office), and the location an employee works at. A question that remains is which type of acts this employee or functionary type performs. We believe the functionary type is the level between employees and actor roles, meaning that V6 must be discarded.

V13: Departments (clustered by e.g. responsibility, competence, market, . . .);
V14: Organizational hierarchy;
V15: Functionary types;
V16: Cross-reference employee/functionary type (replacing V6);
V17: Cross-reference employee/workplace;
V18: Cross-reference functionary type/act type.

There are two more desk officers working in this department.

First, we see an amount of FTE for the functionary type desk officer. Second, the question arises how these persons work together. The variables are:

V19: Amount of FTE (per department, functionary type, location, . . .);
V20: Per act, way of fulfilling actor role (sequentially, concurrently, or collectively [10, p.125]);
V21: Separation of function.

Customer orders are placed through several channels: walk-in, telephone, fax, and e-mail. Walk-in customers are usually people who want to rent a car immediately. Through the other channels one makes in general advance reservations.

V22: Channels, including degree of integration and availability per C-act and workplace.

These can be made up to 200 days in advance.

This is a business rule and is thus present in the ontological model.

In all cases, an electronic rental form is filled out by one of the desk employees, as input to RACES (RAC Information System).

V23: Medium of entering data (writing, typing, voice)

Note that it is the desk employee who registers the request, delegated by the customer, and the promise. Other variables found are:

V24: Medium of gathering data (ask, search on the internet, get from central registrations (external data source), . . .);
V25: Medium of saving data (digitally, paper, human brain, . . .);
V26: Medium of receiving information (sound, image, text, . . .);
V27: Rules for assigning people to tasks;
V28: Order of working;
V29: Language support.

Comparing Table 4 and Table 3, we conclude that our analysis did not reveal many new elementary implementation variables, but mostly implementation variables regarding the coherence between the (elementary) variables from Table 3. For example, the notions "Functionary type" and "Assignment of tasks" were made specific and complemented in cross-references such as employee X functionary type, functionary type X act type, and functionary type X location. Also, all variables and categories from the earlier tables, except for Sourcing, were found in these case descriptions, confirming existing literature.

Table 4. Full list of implementation variables found in EU-Rent case

	Business	Informational	Documental	descr.
Parties and people	Organization structure: organizational/legal entity			1,2
	Employees and Sourcing			1
	Delegation			1,2
	Competences/certification			1
	Addressee specificity[NEW]			1
	Departments			1,2
	Organizational structure			1,2
	Functionary types			1,2
	X-ref employee/functionary type[NEW]			1
	X-ref functionary type/act type[NEW]			1
	#FTE			1
	Way of fulfilling actor role[NEW]			1
	Separation of function			1
	Order of working			1,2
	Assignment of tasks			1
		Language support		2
ICT and other means	Workplaces (including locations of offices)			1,2
	Equipment (including infrastructure)			1
	X-ref workplace/act type[NEW]			1
	Event location restrictions[NEW]			1,2
	Applications (including WFMS, D(BMS), ...)			1
	X-ref employee/workplace[NEW]			1
		Media (entering, gathering, saving, receiving)[NEW]		1
			Channels	1
			Denotation[NEW]	1

5   Conclusions and Future Research

Our ideal was to formulate a list of anticipated changes for which agility is ensured, just like the theory for Normalized Systems [7] did for automated IT systems. This would contribute to uniformity and standardization in the competence of Enterprise Engineers, thus enabling traceability in governing enterprise transformations. Also, where an ontological model gives already a solid starting point for cross-organizationally usable IT applications [22], using the explicit knowledge of organizational implementation variables in an IT platform could turn this into an actual cross-organizationally running IT application. How far have we come with such a list?

First of all, we noticed that our analysis (Section 4) did not reveal new categories, compared to the literature (Section 3), but did reveal implementation variables regarding the coherence between existing variables. For example, the notion Housing (or location) was made specific and complemented in a cross-reference employee X location, act type X location and event location restrictions. Also, the variables and categories from literature were found in and confirmed by the case descriptions.

To have these variables explicit and operationalized offers opportunities for building IT flexibility in a platform. In the example of the student, who is now only allowed to perform drop-offs (accept) and pick-ups (state), it would be possible to present all his tasks related to that in some Google Maps overview with time tables, in which he can also confirm the end of performing a task. Suppose one day RAC decides to allow students to do transports as well, and the IT knows the notion of functionary type X act type etc., then all connected software applications can use this information to change their – e.g., GUI and security – behavior accordingly, potentially without the need for this software to be reprogrammed. Since changing (also organizational) implementation variables tend to have combinatorial effects [7], the future potential for wider validation and application of this list is significant.

We realize this is a modest start on the way to a complete list of organization implementation variables. Therefore we propose the following future research:

- repeat the procedure from Section 4 for real-life observations or procedure descriptions from large organizations;
- add rigor to each variable found: what exactly is its meaning, and why is this variable positioned in a certain EEF-cell;
- validate with existing IT systems to what extent these variables are explicit, and for the implicit variables, where they hinder organizational flexibility;
- elaborate a model for coordination and work flows, including the assignment of subjects to actor roles or functionary types, the assignment of tasks to subjects, and the prioritizing and scheduling of tasks;
- explore functional / constructional gaps, e.g. Quality of Human Services (QoHS) and Quality of Automated Services (QoAS) as functional with respect to Resourcing and IT support respectively.

References

1. Overby, E., Bharadwaj, A., Sambamurthy, V.: Enterprise agility and the enabling role of information technology. Eur. J. Inf. Syst. 15, 120–131 (2006)
2. van Oosterhout, M.P.A.: Business Agility and Information Technology in Service Organizations. PhD thesis, Erasmus University Rotterdam (June 2010)
3. Dietz, J.L.G., Hoogervorst, J.A.P.: Enterprise Ontology and Enterprise Architecture – how to let them evolve into effective complementary notions. GEAO Journal of Enterprise Architecture 1 (2007)
4. Conboy, K., Fitzgerald, B.: Toward a Conceptual Framework of Agile Methods: A Study of Agility in Different Disciplines. In: Proceedings of the 2004 ACM Workshop on Interdisciplinary Software Engineering Research, WISER 2004, pp. 37–44. ACM, New York (2004)
5. Sarkis, J.: Benchmarking for agility. Benchmarking: An International Journal 8(2), 88–107 (2001)
6. Seo, D., La Paz, A.I.: Exploring the Dark Side of IS in Achieving Organizational Agility. Commun. ACM 51(11), 136–139 (2008)

7. Mannaert, H., Verelst, J.: Normalized Systems: Re-creating Information Technology Based on Laws for Software Evolvability, Koppa, Kermt, Belgium (2009)
8. Object Management Group: Business Motivation Model (BMM) Specification, V1.1. OMG Available Specification OMG Document Number: formal/2010-05-01 (May 2010), http://www.omg.org/spec/BMM/1.1/PDF/
9. Op 't Land, M., Dietz, J.L.G.: Benefits of Enterprise Ontology in Governing Complex Enterprise Transformations. In: Albani, A., Aveiro, D., Barjis, J. (eds.) EEWC 2012. LNBIP, vol. 110, pp. 77–92. Springer, Heidelberg (2012)
10. Dietz, J.L.G.: Enterprise Ontology – Theory and methodology. Springer (2006)
11. Dietz, J.L.G.: Architecture: Building strategy into design. Sdu Uitgevers bv, The Hague, The Netherlands (2008)
12. van Dipten, E., Mulder, J.B.F.: Basic Enterprise Engineering Map. Informatie 10, 54–61 (2011)
13. Op 't Land, M., Proper, H.A.: Impact of Principles on Enterprise Engineering. In: Österle, H., Schelp, J., Winter, R. (eds.) Proceedings of the 15th European Conference on Information Systems, pp. 1965–1976 (2007)
14. Op 't Land, M., Pombinho, J.: Strengthening the Foundations Underlying the Enterprise Engineering Manifesto. In: Albani, A., Aveiro, D., Barjis, J. (eds.) EEWC 2012. LNBIP, vol. 110, pp. 1–14. Springer, Heidelberg (2012)
15. Tsourveloudis, N.C., Valavanis, K.P.: On the Measurement of Enterprise Agility. Journal of Intelligent and Robotic Systems (33), 329–342 (2002)
16. Sherehiy, B., Karwowski, W., Layer, J.K.: A review of enterprise agility: Concepts, frameworks, and attributes. International Journal of Industrial Ergonomics 37(5), 445–460 (2007)
17. BIZZdesign: Handboek Business Process Engineering. Academic version 7.1 (in Dutch) edn. BIZZdesign B.V. Academy Publishers (2009)
18. Hoogervorst, J.A.P.: A framework for enterprise engineering. International Journal of Internet and Enterprise Management 7(1), 5–40 (2011)
19. de Bruin, B., Verschut, A., Wierstra, E.: Systematic Analysis of Business Processes. Knowledge & Process Management 7(2), 87–96 (2000)
20. Op 't Land, M.: Applying Architecture and Ontology to the Splitting and Allying of Enterprises. PhD thesis, Delft University of Technology (2008)
21. Rychkova, I.: Towards Automated Support for Case Management Processes with Declarative Configurable Specifications. In: La Rosa, M., Soffer, P. (eds.) BPM Workshops 2012. LNBIP, vol. 132, pp. 65–76. Springer, Heidelberg (2013)
22. Krouwel, M., Op 't Land, M.: Using Enterprise Ontology as a basis for Requirements for Cross-Organizationally Usable Applications. In: Figueiredo, A.D., Ramos, I., Trauth, E. (eds.) Proceedings of the 7th Mediterranean Conference on Information Systems 2012 (MCIS 2012). MCIS Proceedings, University of Minho, Portugal, AIS Electronic Library (AISeL), Paper 23 (2012)
23. Scheithauer, G., Hellmann, S.: Analysis and Documentation of Knowledge-Intensive Processes. In: La Rosa, M., Soffer, P. (eds.) BPM Workshops 2012. LNBIP, vol. 132, pp. 3–11. Springer, Heidelberg (2013)
24. Dietz, J.L.G.: The Essence of Organization - an introduction to Enterprise Engineering. Sapio (2013), to be published @ http://www.demo.nl
25. Schacher, M.: Mini EU-Rent: Business Model. Technical Report v23.06.2008, KnowGravity (2008),
http://www.knowgravity.com/pdf-e/Mini%20EU-Rent%20BU.pdf

A Case Study on Enterprise Transformation in a Medium-Size Japanese IT Service Provider: Business Process Change from the Ontological Perspective

Sanetake Nagayoshi

Department of Industrial Engineering and Management,
Graduate School of Decision Science and Technology, Tokyo Institute of Technology
2-12-1, Ookayama, Meguro-Ku, Tokyo, 152-8550, Japan
nagayoshi.s.aa@m.titech.ac.jp

Abstract. Organizational change and transformation is an important research topic that attracts the attention of many researchers and practitioners in organization research. This paper describes several case studies of applying DEMO in enterprise transformation. The enterprise transformation of divisions in Company A is analyzed to examine the relationship between Product-Market Growth Grid and business process change from the ontological perspective. The results indicate that (1) it is not always necessary to change the ontological level of business process in market development, (2) it is necessary to change the ontological level of business process in product development, and (3) it is necessary to reengineer the ontological level of business process in diversification. The generalizability of these results can be ascertained with more studies in future.

Keywords: Business Process Change, Enterprise Transformation, Product-Market Grid, Ontological View, DEMO.

1   Introduction

Organizational change and transformation is an important research topic that has attracted the attention of many researchers and practitioners in organization research. Traditionally, organizational change and transformation mainly focuses on the change in organizations' structures and practices. However, as organizations increasingly face changes in the economical, geopolitical, contextual, and technological aspects, discourses on organizational transformation has begun to draw from many other disciplines and various perspectives. For example, studies of technology-led organizational transformation obtain insights from both information systems and organization research. As a result, a plethora of studies has evolved in this arena, with emphasis on the use of process-based research and rhetoric approaches in understanding how and why organizations change and transform.

H.A. Proper, D. Aveiro, and K. Gaaloul (Eds.): EEWC 2013, LNBIP 146, pp. 43–57, 2013.
© Springer-Verlag Berlin Heidelberg 2013

The objective of this paper is to describe business process change based on case studies of enterprise transformation from the ontological perspective. In this paper, the relationship between market diversification (in the Product-Market Growth Grid) [1] (Figure 1) and business process change from the ontological perspective is analyzed and discussed, and the business process diagrams based on DEMO (Design and Engineering Methodology for Organization) are described [2].

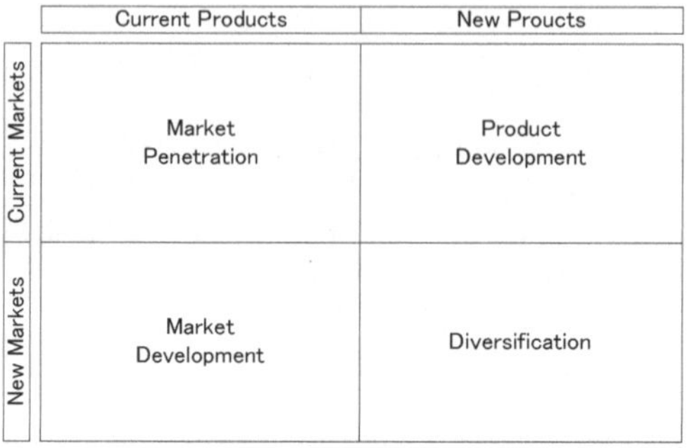

	Current Products	New Proucts
Current Markets	Market Penetration	Product Development
New Markets	Market Development	Diversification

Fig. 1. Product-Market Grid [1]

The remainder of this paper is organized as follows: First, related studies are reviewed in section 2. The research method of this study is described in section 3. Next, five cases of transformation in a Japanese IT service provider are introduced in section 4. Then, the cases are analyzed in section 5 and discussed in section 6. Finally, the limitations of this study and implications for future research are described in section 7.

2   Literature Review

Enterprises increasingly need to consider and pursue fundamental change, such as upgrading current business and innovating in terms of expansion, M&A, and globalization, in order to maintain or gain competitive advantage. To distinguish from traditional routine changes, fundamental change is referred to as enterprise transformation [3].

2.1   Business Process Change

Enterprise transformation is enabled by *"work process change"*, which requires the *"allocating of attention and resource"* so that an enterprise can anticipate and adapt to changes with their resources to yield improvement [3]. Enterprise transformation is a more innovative and strategic change that influences multiple aspects of an organization, including routine, organization structure, human capital, and marketing strategy. Compared with low-level changes, it is more difficult to model and manage [4] [5]. Miles, Snow, Meyer and Coleman [6] proposed a theoretical framework that deals with alternative ways in which organizations define their product-market domains (strategy) and construct mechanisms (structures and processes) to pursue the strategies. Hamel and Prahalad [7] examined why many companies disbanded or dramatically downsized their strategic planning departments. This led to focus on the concept of "business process re-engineering" proposed by researchers such as Hammer and Champy [8], which involves deep redesign of business processes. The concept was popular during the 1990s as a reaction to recession, during which companies needed to downsize and better apply information technology [9] [10] [11]. Business process reengineering seeks to achieve dramatic performance improvement by radically redesigning an organization and its takes precedence over information systems development which has focused mainly on automating and supporting existing organizational procedures [12].

Mintzberg et al. [13] pointed out that Wernerfelt [14] proposed the Resource-based Theory, although Wernerfelt [15] himself maintains that his ideas were not popular until Prahalad and Hamel [16] developed the concept of dynamic capabilities. Firms can be thought of as collections or accretions of resources, which are heterogeneously distributed and persist over time. When firms possess resources that are valuable, rare, inimitable, and non-substitutable, they can achieve sustainable competitive advantage [17] in a turbulent environment.

Nagayoshi, Liu and Iijima [18] argued that exception handling in business processes can be classified into eight patterns from the language actor perspective, which can trigger enterprise transformation.

2.2 Business Rule Management

Ansoff [1] and Drucker [19] believe that planning is an integral part of a well-managed company. Andrews [20] discussed the functions of general management, particularly the role of the chief executive and senior vice-presidents. He believed that in addition to maintaining surveillance over the actual attainment of results formally or informally planned, the general manager should be expected to make, or at least preside over, the process of creating policy decisions that will affect future results.

Different research streams have emerged to provide various insights ranging from business rule authoring, engineering, rule mining, and many others. A business rule refers to "a statement that defines or constrains some aspect of the business. This must be a term or fact (described as a structural assertion), a constraint (described as an action assertion), or a derivation. It is 'atomic' in that it cannot be broken down or decomposed further into more detailed business rules. If reduced any further, there would be loss of important information about the business." Halle [21] defined the business rule approach to systems development as one that "allows the business to automate its own intelligent logic better, as well as to introduce change from within itself and learn better and faster to reach its goals."

Maglio, Srinivasan, Kreulen and Spohrer [22] mentioned that we can view business rule management from a service science perspective, where service systems are defined as "value-creation networks composed of people, technology and organizations."

2.3   Organization Learning

Enterprise transformation is related to organization learning, which is a process of detecting and correcting errors [23] to improve an organization's value creation capabilities.

Other knowledge management processes such as SECI [24], along with cultural changes and structural changes, can increase an enterprise's "readiness for transformation". They are therefore likely to enable a successful transformation.

Slater and Narver [25] described the processes through which organizations develop and use new knowledge to improve performance.

Liu, Nagayoshi and Iijima [26] discussed how enterprise transformation, business process change, and organization learning affect each other. They identified "readiness" to be an important factor for successful enterprise transformation. This factor could be affected by organization learning, organizational culture, and organizational structure. Liu, Nagayoshi and Iijima [26] also provided a promising way of analyzing enterprise transformation or innovation-related problems at the ontological level from an engineering perspective, which can greatly reduce the complexity of problems while significantly enhancing effectiveness and efficiency.

2.4   Marketing Strategy

Marketing strategy also interacts with enterprise transformation.

Product-Market Growth Grid [1] is a widely adopted framework that conceptualizes market diversification, which is a kind of marketing strategy. It includes market penetration, market development, product development, and diversification.

The framework is intuitive and some longitudinal descriptive studies, such as that by Miller and Friesen [27], have provided some qualitative evidence for aspects of the framework. Greiner's model [28] provides some causal explanation by hypothesizing that growth occurs in relatively stable phases, interspersed with "crises". During a crisis, an organization either successfully adapts or fails. In this sense, crises may be seen as necessary catalysts of learning and further growth.

Although enterprise transformation is interpreted from various points of view nowadays, "how" and "why" enterprises change and transform remain critical research topics. Accordingly, this paper examines marketing strategy to explain "why" organizations change and business processes change to describe "how" organizations change.

Also, to the best of my knowledge, there has been a lack of academic studies on how enterprise transformation, marketing strategy, and business processes at the ontological perspective affect each other.

3   Research Method

3.1   Engineering Perspective

Given the complexity of change, several researchers have argued that enterprise transformation should be considered from the modular and system perspectives rather than the workflow perspective. After a business is conceptualized and modularized, similar businesses can be managed using the same dominant logic [29]. This can help to reduce the complexity of change and improve analyzability. Op't Land and Dietz [5] suggested that DEMO (Design and Engineering Methodology for Organization) [2], an enterprise ontology, can be used for modeling enterprise transformation while greatly reducing the return on modeling effect (ROME).

DEMO is an enterprise ontology used to describe the essential structure of an enterprise without getting into implementation details. It includes four aspect models: Construction Model, Process Model, Action Model, and State Model [2]. The interaction model (IAM) is included in the construction model, as shown in Figure 2 and it aims to describe the construction of an enterprise by defining:

- **Transaction:** A sequence of acts between two actor roles, including communication loop acts (request, promise, state, and accept) and production acts.

- **Actor Role:** An actor role is defined in terms of responsibility, authority, and capability. As shown in Figure 2, A01 is an actor role which initiates the transaction T02, and A02 is the actor role that executes the transaction (T02).

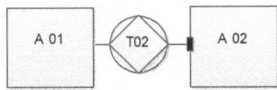

Fig. 2. ATD (Actor Transaction Diagram)

- **Transaction and Result Table:** As shown in Table 1, each transaction has a result which expresses how the transaction changes the state of a given context.

Table 1. TRT (Transaction Result Table)

Transaction	Result
T02 Prepare	R02 Pizza P is prepared

By using the DEMO IAM model, the main structure of an organization as well as how its main business is conducted, who is involeved, what responsibilities they take, and what the results are can be clarified totally without getting into implementation details.

Several studies have applied DEMO. Aveiro [30] discussed how to automatically detect and address dysfunctions in exception handling using DEMO from an engineering viewpoint. Further, Nagayoshi, Liu & Iijima[18] developed patterns for exception handling, considering not just routine changes but also structural changes and action rule changes. From the value creation viewpoint, Pombinho[31] linked the construction of an enterprise with the concept of value to examine "whether we do the right thing" by analyzing the value stream.

DEMO IAM is used as an enterprise transformation description method in this paper, because it describes the essential structure of an enterprise without too much details.

3.2   Qualitative Research

Adopting the engineering perspective, a qualitative study was conducted. The case study was based on four interviews with several managers of "Company A" from April to June 2012:

- **First Interview:** Director of Company A,

- **Second Interview:** CEO and technical director of Company A,

- **Third Interview:** Sales manager and technology manager from the second solution division,

- **Fourth Interview:** Sales manager and technology manager from the first solution division.

Each interview lasted for 2 to 3 hours. The interviews are analyzed and the results are described in the following sections. To ensure the accuracy of interpretations, the results were reviewed by the director of Company A.

4   Case Studies

Company A is a Japanese IT service provider founded in 1969 as a software provider. After a long period of growth, the company designed and developed several well-known application packages in the early 1980's. One of them is an accounting system for the local government. It was awarded for being an "outstanding information system". "First division" of the company provides system integration services mainly for the local government in Japan by implementing their own software packages.

Company A expanded its businesses into system integration from the late 1980's. In the business, Company A takes the role of a sub-constructor and dispatches work to users' companies according to primary constructors' solutions. Primary constructors' generate solutions by identifying and analyzing customer requirements.

With the evolution of new technology (cloud computing, concrete secure technology, etc.) and keen competition in information technology, Company A had to identify new business opportunities to enhance their competitiveness. The transformation inside Company A includes the following five transformations in the business logic of the following solution divisions:

- **First Division:** First division made one business logic change, which was to transform from **a software package provider** to **an application service provider** (software as a service type of business).

- **Second Division:** Second division made two business logic changes. They were (1) delivery change from a **passive-type business** to an **active-type business**, and (2) organizational change to conduct the active-type business.

- **Third Division:** Company A acquired a business of application software package for managing IT security from an existing company. The package had various functions such as single sign-on, authentication with IC card and secure printing. Company A reorganized the third division for the new software package business. The division made two business logic changes: (1) When "Third Division" **started the new business**, the software package was modified to serve their own existing customers, and (2) Based on the new software package, the division found **new market opportunities and new customers** in other industries.

The five transformations are described in the following sections. The description for each case of transformation is based on DEMO IAM. Each case is analyzed from the construction change aspect.

5 Case Analysis

5.1 First Division

A. From a software package provider to an application service provider (software as a service type of business). This is a transformation in the First Division of Company A, which had focused on "providing software package with installation" before the transformation.

Providing support to the local government was a traditional business of the First Division of Company A. As a package provider, their main targets were small and medium cities and towns in Japan, which helped to distinguish Company A from the other e-government IT service providers.

For Company A to be competitive, it needed to transform from being a package provider to a service provider.

Company A made the first step by successfully providing application service to "regional government D". Primary challenges, transformations and solutions were identified as follows.

The original construction model of the First Division as a provider of software package with installation is shown in Figure 3.

In order to provide application services, it was necessary to consider risk control. In the transformed business, sales persons need to estimate costs and risks for providing a service before proposing the service to customers. As a service provider, Company A needs to consider the risks and keep them under control. Therefore, a new transaction T11 was added, as shown in Figure 4.

There were also some changes to transaction T05. T05 was no longer initiated by the order completer. Instead, it was initiated by the product manager, so that a "work plan" could be made before an order was received. T04 and T07 remained the same but the action rules changed. For transaction T04, Company A supplied a service to customers instead of shipping the software package to customers. Accordingly, the rule of T07 "payment" changes from being based on the price of a software package to charging customers for services.

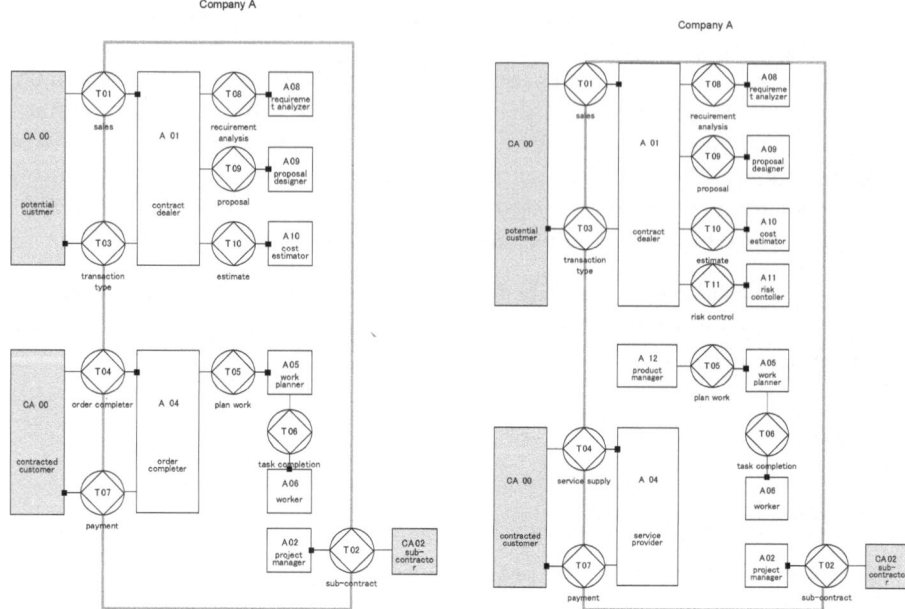

Fig. 3. ATD before in First Division   **Fig. 4.** ATD after in First Division

5.2   Second Division

B-(1). From a passive-type business to an active-type business. Before transformation, Company A mainly played the role of a sub-constructor. The primary constructors were the ones who had the "know-how" knowledge to provide solutions for fulfilling customers' requirements. Sub-constructor only focused on dispatching skilled workers (who have different unit prices depending on their skill levels) to customers to support software development according to primary constructors' requirements. In this type of business, a sub-constructor does not need to have much knowledge about a customer's business. Also, it is not necessary to manage the schedule and risk of a software development project as these are the responsibilities of the primary constructor. The revenue from dispatching workers was calculated by multiplying unit price and work hour. The construction model of this business is shown in Figure 5.

Company A conducted this type of business for about 10 years. "Company B" was one of the biggest customers of Company A, which was one of the biggest Japanese IT infrastructure and systems integrators. Company A accumulated plenty of experience not only in software development but also in different business sectors, especially in billing systems. These had driven its transformation from being a "sub-constructor" to a "primary constructor".

As a primary constructor, Company A provided another type of contract choice - Request for Proposal. In the transformed business, Company A was responsible for delivering the final solution while controlling for time, quality, and fixed cost.

The whole solution included: requirement analysis, architecture design, software development, and testing.

The construction model of the transformed business is shown in Figure 6.

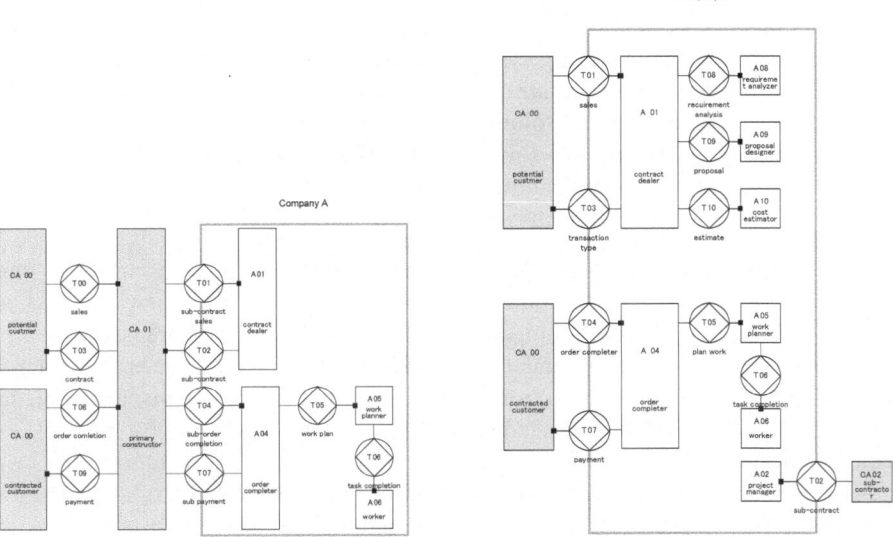

Fig. 5. ATD Sub Constructor   **Fig. 6.** ATD Prime Contractor

The first success of this new business is "Project C" in year 2009.

"Company C" was a traditional distributor in Japan. In 2009, Company C planned to offer a new service product and they needed an effective billing solution to support the new service product. Company A was chosen as the solution provider. This project lasted from October 2010 to October 2011.

Based on interviews of the project manager and the chief sales person in the Second Division, we identified several changes as discussed next.

As shown in Figure 5, Company A did not need to propose solutions in its old business. The sales person who played the actor role "A01" passively sold human resources in terms of skilled workers dispatched to customers. Based on primary constructor's requirements, sales person provided information about their staff and made contract with a customer by counting the number of workers by hour. Clearly, transactions T00, T03, T06, and T09 were not within the business of Company A before the transformation.

After the transformation, T00, T03, T06, and T09 in Figure 5 became part of the business of Company A. A proposal group rather than a sales person was involved in the sales process. As shown in Figure 6, three new transactions (T08, T09, and T10) are added as sub-transactions of transaction T01, which indicates that Company A needs to analyze customers' requirements, make proposals to customers, and estimate costs in order to complete "pre-sales". A sales process is complete and a contract can be made only when all these transactions are finished.

The actor roles (i.e., A08, A09, and A10) corresponding to each transaction also changed in terms of the requirements for capability, responsibility, and authority. A proposal group includes:

- A pre-sales person who plays the role of A08 as a "requirement analyzer", with strong capability to understand customers' requirements.
- A pre-sales person who plays the role of A10 as a "cost estimator", with the responsibility of estimating costs and negotiating prices with customers based on the proposed solution.
- Consulting staff who plays the role of A09 as a "proposal designer", with the capability of understanding technical details as well as designing and proposing solutions to customers.

In the old business of Company A, the primary constructor or the customer must control the cost, quality, and schedule of a project. However, in the new business, all these responsibilities lie with Company A. In Figure 6, although transactions T01, T04, T05, T06, and T07 have similar construction with those in Figure 5, the action rules for the corresponding actor roles had changed. For project management, the content of T05 "work plan" expanded to include items such as cost control and quality control.

As a primary constructor, Company A also delegate work to sub-constructors according to their work plan. Thereforel, transaction T02 "sub-construction" can be initiated by a project manager in Company A.

B-(2). Organizational change to conduct primary-constructor-type business. To conduct the new business, it is necessary for pre-sales persons playing the actor roles of A08 and A10 to acquire additional communication skills, proposal skills, and "know-how" knowledge. Company A did not have such skills and knowledge in its old business. However, as it became a subsidiary company of a large IT service provider, Company A was able to hire several experienced sales persons with relevant pre-sales skills and knowledge. This helped the sales persons in Company A to obtain more skills and knowledge about pre-sales.

A "proposal group" was also formed in Company A. It included pre-sales persons, managers, and consultants who played the actor roles of A01, A08, A09, and A10. The "proposal group" works closely with customers to understand their business and requirements and proposes suitable solutions. The group needs to carefully estimate and manage costs, risks, and schedule. When a proposal is accepted by a customer, the analysis and design process will continue until the delivered solution is acknowledged by the customer.

Project management became a big challenge for managers in the Second Division. They had to acquire the ability to manage projects through practical experience and on-the-job training.

However, from the construction-change point of view, the ATD for primary-constructor-type business is the same ATD in figure 6.

5.3 Third Division

C-(1). Software package modification to start a new software package business.
Before acquiring the new application software for managing IT security and forming
the Third Division, Company A had several application software packages. The
business processes for the old business is described in ATD in Figure 7. After
acquiring the new application software, the Third Division needed to modify it
according to the brand image of Company A and to cater to the needs of their existing
customers. The business processes for the modification and sales are described in
ATD in Figure 8.

Comparing Figure 7 and Figure 8, it can be observed that the transactions T10,
T11, T12 and T13 and actor roles CA01, A09, A10, and A11 in the sale of application
software are similar. However, it was necessary for Company A to add some
important activities such as market investigation, market requirement analysis, cost
estimation, risk assessment, decision making for investment, and modification in
order to sell the new application software. These additional transactions and actor
roles are shown as T01-T08 and A01-A08 respectively in Figure 8.

C-(2). Applying the software package to a new market. As many industries had
strong need for IT security, the Third Division had the opportunity to expand its
market. Since IT security requirements are similar in most industries, the Third
Division was able to sell the new software to many industries without much
modification.

However, from the construction-change point of view, the ATD for selling the
new software package to the new customer in the new industry markets is the same
ATD in figure 7.

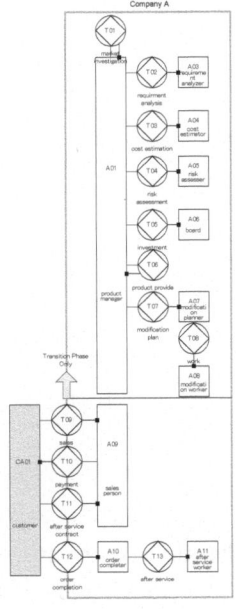

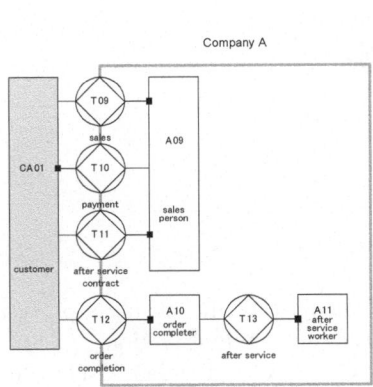

Fig. 7. ATD Sub-Constructor   **Fig. 8.** ATD Primary Constructor

6   Discussion

The cases of enterprise transformation in Company A can be classified into three categories based on the Product-Market Growth Grid [1].

Case A in the First Division is a kind of product development, in which a software package provider develops into an application service provider to serve the local governments in Japan.

Case B-(1) in the Second Division is a kind of diversification. For the product aspect, Company A transformed from a sub-constructor providing worker dispatching services to a primary constructor providing systems integration services. For the market aspect, Company A transformed from serving customers such as system integrators to serving new customers in the distribution industry.

Case B-(2) in the Second Division is a kind of market development, which involves providing existing system integration service as a primary constructor not only to the current customers in the distribution industry but also to other new industries.

Case C-(1) in the Third Division is a kind of product development, in which Company A acquired an application package from another company and modified it in order to provide IT security services to existing customers.

Case C-(2) in the Third Division is a kind of market development, which involves providing existing (acquired and modified) application software for managing IT security not only to the existing customers but also new customers in other industries.

These transformations are illustrated in Figure 9.

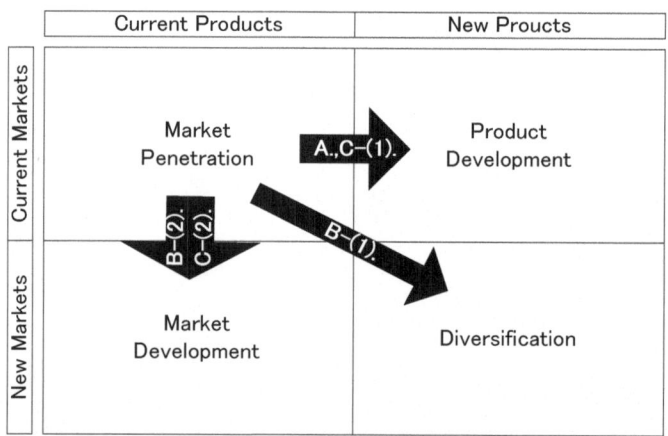

Fig. 9. Transformations in "Company A"

Next, these transformations will be considered along with the business process changes in each case, which were described in ATDs in the previous section.

In the cases of product development like Case A and Case C-(1), the transformations involved business process changes. Examples include the additional internal transactions and internal (elementary) actor roles as described in Figure 3 and 4 for Case A and in Figure 7 and 8 for Case C-(1).

In the cases of market development like Case B-(2) and C-(2), the transformations involved few business process changes.

In the case of diversification like B-(1), the transformation involved structural business process changes. For example, internal transaction and internal (elementary) actor roles were added, and external transaction and external (composite) actor roles were not only added but also eliminated as described in Figure 5 and 6.

Conclusion. According to the case studies in Company A, it is observed that (1) it is not always necessary to change the ontological level of business processes in the case of market development, (2) it is necessary to change the ontological level of business processes in the case of product development, and (3) it is necessary to reengineer the ontological level of business ontological processes in the case of diversification.

When a company intends to sell its existing products to new customers, change in business processes may not be necessary, because the company only needs to know customer information such as implicit and/or explicit customer needs and sometimes industry-specific knowledge. This requires substantial changes in employee behavior, company culture, and information systems which are not observable at the ontological level. This suggests that an ontological-level analysis may miss out some important changes. When a company intends to deal with new products and/or services, it often needs to change business processes because production and/or service delivery is directly related with business processes.

7   Conclusion

In this paper, enterprise transformations in Company A were analyzed based on the marketing diversification aspect of the Product-Market Growth Grid [1] and ontological business process changes. This study demonstrates real-world examples of applying DEMO to study transformations. The key findings are:

(1) It is not always necessary to change the ontological level of business processes in the case of market development,
(2) It is necessary to change the ontological level of business processes in the case of product development, and
(3) It is necessary to reengineer the ontological level of business ontological processes in the case of diversification.

These results suggest that it is fruitful for academic researchers to dedicate themselves to further study the relationship between the marketing diversification aspect of the Product-Market Growth Grid [1] and business process changes.

This paper did not discuss market penetration. Changes in business processes are not expected except when an organization aims to improve business processes. However, further evidence is needed to support this.

This study is mainly qualitative and it is necessary to collect more quantitative evidence to further support the findings. To this end, it is also necessary to develop quantitative indicators. More studies are needed to generalize the findings beyond this study.

Acknowledgement. The author thanks the CEO, the director, the senior manager, the sales managers and the technology managers in Company A, who kindly dedicated time to my research work. I could not achieve my work without their generous cooperation.

References

1. Ansoff, I.: Corporate Strategy. McGraw-Hill, New York (1965)
2. Dietz, J.: Enterprise Ontology: Theory and Methodology. Springer-Verlag New York Inc., New York (2006)
3. Rouse, W.B.: A theory of enterprise transformation. Systems Engineering 8(4) (2005)
4. Op 't Land, M., Proper, E., Waage, M., Cloo, J., Steghuis, C.: Enterprise architecture: Creating value by informed governance. Springer (2009)
5. Op 't Land, M., Dietz, J.L.G.: Benefits of enterprise ontology in governing complex enterprise transformations. In: Albani, A., Aveiro, D., Barjis, J. (eds.) EEWC 2012. LNBIP, vol. 110, pp. 77–92. Springer, Heidelberg (2012)
6. Miles, R., Snow, C., Meyer, A., Coleman Jr., H.: Organizational Strategy, Structure, and Process. The Academy of Management Review 3(3), 546–562 (1978)
7. Hamel, G., Prahalad, C.K.: Thinking differently. Business Quarterly 59(4), 22–35 (1995)
8. Hammer, M., Champy, J.: Re-engineering the Corporation: A Manifesto for a Business Revolution. Harper, New York (1993)
9. Davenport, T.H., Short, J.E.: The new industrial engineering: Information technology and business process redesign. Sloan Management Review 31(4), 11–27 (1990)
10. Cole, R.: Reengineering the corporation: A review essay. Quality Management Journal 1(4), 77–85 (1994)
11. Mumford, E.: New treatments or old remedies: Is business process reengineering really socio-technical design? Journal of Strategic Information Systems 3(4), 313–326 (1994)
12. Guha, S., Kettinger, W., Teng, J.: Business Process Reengineering: Building a Comprehensive Methodology. Information Systems Management 10(3), 13–22 (1993)
13. Mintzberg, H., Ahlstrand, B., Lampel, J.: Strategy Safari: A Guided Tour through the Wilds of Strategic Management. Prentice-Hall, New York (1998)
14. Wernerfelt, B.: A resource-based view of the firm. Strategic Management Journal 5(2), 171–180 (1984)
15. Wernerfelt, B.: The resource-based view of the firm: ten years after. Strategic Management Journal 16(3), 171–174 (1995)
16. Prahalad, C.K., Hamel, G.: The core competence of the corporation. Harvard Business Review 68(3), 79–91 (1990)
17. Teece, D.J., Pisano, G., Shuen, A.: Dynamic capabilities and strategic management. Strategic Management Journal 18(7), 509–533 (1997)
18. Nagayoshi, S., Liu, Y., Iijima, J.: A study of the patterns for reducing exceptions and improving business process flexibility. In: Albani, A., Aveiro, D., Barjis, J. (eds.) EEWC 2012. LNBIP, vol. 110, pp. 61–76. Springer, Heidelberg (2012)

19. Drucker, P.: Long-range planning. Management Science 13(2), 238–249 (1959)
20. Andrews, K.R.: The Concept of Corporate Strategy. Dow Jones-Irwin, Homewood (1987)
21. von Halle, B.: Business Rules Applied: Building Better Systems Using the Business Rules Approach. John Wiley & Sons, Inc., New York (2002)
22. Maglio, P., Srinivasan, Kreulen, J.T., Spohrer, J.: Service systems, service scientists, SSME and innovation. Communications of the ACM 49(7), 81–85 (2006)
23. Fiol, C.M., Lyles, M.A.: Organizational learning. The Academy of Management Review 10(4), 803–813 (1985)
24. Nonaka, I., Takeuchi, H.: The Knowledge-Creating Company: How Japanese Companies Create the Dynamics of Innovation. Oxford University Press, New York (1995)
25. Slater, S.F., Narver, J.C.: Market Orientation and the Learning Organization. Journal of Marketing 59(3), 63–74 (1995)
26. Liu, Y., Nagayoshi, S., Iijima, J.: Innovative Transformation In a Knowledge Intensive Industry: A Case Study of an Organizational Learning Based Enterprise. In: International Conference on Inclusive Innovation and Innovative Management (2012)
27. Miller, A., Friesen, P.: A longitudinal study of the corporate life cycle. Management Science 30, 1161–1183 (1984)
28. Greiner, L.E.: Evolution and revolution as organizations grow. Harvard Business Review (July/August 1972)
29. Bettis, R.A., Prahalad, C.K.: The Dominant logic: retrospective and extension. Strategic Management (16) (1995)
30. Aveiro, D.S.: G.O.D. (Generation, Operationalization & Discontinuation) and Control (sub)organizations: A DEMO based approach for continuous real-time management of organizational change caused by exceptions. UTL, Lisbon (2010)
31. Pombinho, J., Tribolet, J.: Service system design and engineering – A value-oriented approach based on DEMO. In: Snene, M. (ed.) IESS 2012. LNBIP, vol. 103, pp. 243–257. Springer, Heidelberg (2012)

Explaining with Mechanisms
and Its Impact on Organisational Diagnosis

Roland Ettema[1], Federica Russo[2], and Philip Huysmans[3]

[1] TU Delft, Jaffalaan 5, 2628 BX Delft, Netherlands
roland.ettema@gmail.com
[2] Center Leo Apostel, VU Brussel, Krijgskundestraat 33, B-1160 Brussels, Belgium
ferusso@vub.ac.be
[3] University of Antwerp, Prinsstraat 13, 2000 Antwerp, Belgium
philip.huysmans@ua.ac.be

Abstract. Lean Six Sigma (LSS) is the leading approach in organizational diagnosis. This approach is largely based on the analysis of correlations (e.g., Multivariate Analysis (MANOVA)), which constitutes the main source of information to establish the causes of dysfunction and to indicate possible interventions to restore good functioning. In this paper, we argue that causal mechanisms (CMs) should also be integrated into LSS for organizational diagnosis (OD). We borrow the concept of CM from the field of causality in the sciences. CMs have the potential to improve diagnostic practice because they reveal the structure and the functioning of an organization, and thus indicate more clearly how to intervene in order to restore good functioning. While the LSS movement has been enormously successful in advancing our diagnostic practices, further improvement is possible once causal mechanisms are brought into the picture.

Keywords: Organizational Diagnosis, Causal Mechanism, Lean Six Sigma, ArchiMate, Enterprise Ontology.

1 Introduction

Organization Diagnostics (OD) is the field of study that deals with finding explanations for quality problems in business processes [13, 1, 26, 25, 2]. It is derived from the fields of quality analysis and industrial statistics. The objective of OD is to gain insight into the possible causes of a quality problem on the basis of observing and analyzing statistical data. In other words, statistical correlations are used to diagnose organizational problems. Recently, authors from the social sciences argued that explaining a certain phenomenon using a *causal mechanism* (CM) makes the explanation more intelligible and understandable [6, 5, 21, 20]. However, this claim has neither been investigated in management science, nor introduced in OD [13, 1, 26, 25, 2]. This article introduces the concept of causal mechanism in organizational diagnosis by presenting a case study in which we describe how OD works in practice and explore how the diagnostic process can

H.A. Proper, D. Aveiro, and K. Gaaloul (Eds.): EEWC 2013, LNBIP 146, pp. 58–72, 2013.

be improved by integrating evidence of mechanisms. From an academic perspective, this analysis aims to clarify whether the claim from social sciences and the mechanisms literature (i.e., that explaining a phenomenon on the basis of CM is more intelligible and understandable) also holds true in OD. From a practical perspective, this analysis could lead to a more effective OD practice.

We start this paper by presenting how CMs are defined in the philosophy of science and social sciences. CMs trace back to the pioneering works of Bechtel [3] and Machamer, Darden and Craver [17], and were further improved by Illari and Williamson [16]. While different definitions exist for CMs, it is widely accepted in different fields that an explanation based on CM contains the following components: (1) a mechanism is responsible for a phenomenon; (2) mechanisms consist of entities and activities; and (3) the organization and operational conditions must be made explicit. To apply CMs in organizational diagnosis, these three aspects must be addressed.

To assess the extent to which organizational diagnosis already adheres to these aspects, we will describe in detail how an organizational diagnostic project is performed. We expect to demonstrate that the current state of affairs in OD fails to use CMs in their diagnosing results. We will, therefore, review a case study in two stages. Firstly, we will analyze how the project was performed from the perspective of CMs. Secondly, we will explore which elements could be added to obtain a more complete specification of a CM.

This paper presents the results of the first stage of a larger research project, aimed at answering two questions: (I) What are the consequences for OD if the identification of CM is adopted as its primary objective? To answer this question, an approach must be formulated which addresses the three aspects of CM discussed in this paper. We will argue that the current approach of OD (i.e., relying on statistical analysis and functional modelling) cannot adequately describe the entities and activities of CMs, and that, therefore, the research project should also find an answer to another question: (II) What ontology for business processes can help determine the cause(s) of a quality problem?

The paper is organised as follows. in Section 2 we study the subject of CMs and causality from the point of philosophy of sciences. In Section 3 we describe an OD case study where experts endeavour to improve the quality of a business process by applying statistics and functional modelling. In Section 4 a reflection on the current state of affairs in OD is presented, built on the characterisation of a CM from Section 3 and the case study description. Finally, in Section 5 we present our conclusions and an outline for future research.

2 Explaining with Mechanisms

Explanation has a long tradition in philosophy of science. A notable characteristic of this tradition has been to develop theories of explanation tailored to the natural sciences, and especially physics. For instance, the *deductive-nomological* (D-N) model developed by Hempel and Oppenheim [14] requires that, to explain the occurrence of a phenomenon, it must be deduced from general laws of nature and initial conditions. Later on, Salmon argued that the identification of the basic (physical) processes provides the explanation of the phenomenon at stake [22, 23]. Consequently, the D-N model was abandoned for a theory that put causality prominently into the explanation, but still with physics as its main scope of applicability. As a result, this account was considered unsatisfactory by [17], because such processes do not fit biology or the social sciences [20]. Thus, a 'new turn' in philosophy of explanation started, investigating the importance of CM in explanatory practices.

Social scientists – like many other scientists – face the problem of explaining correlations amongst variables of interest. As is well known, correlations are a fallible source for causation of explanation, hence the need to ground explanations on CMs [24]. Consider for example the relation between smoking and cancer. While correlations between the two have been known for at least two hundred years, it took a long time to conclusively establish a causal relation based on a CM that was clearly identifiable, non-contingent, and stable across time and populations [20, 21]. This search for CMs indicates that social scientists were not satisfied with co-variation between variables only but were also interested to reveal why and through what pathways the outcome was actually generated [20, 18]. So, social scientists are not satisfied with a "black box" approach. After two decades the debate on the right characterisation of a CM, based on Bechtel and Richardson's 1993 book [3], has once again been intensified by Machamer, Darden and Craver's thought-provoking 2000 paper [17]. In this paper, we will adopt a definition from [16] that aims to provide an understanding of what is common to CMs across several fields:

> ' A **mechanism** for a phenomenon consists of entities and activities organized in such a way that they are responsible for the phenomenon.'
>
> *(Illari and Williamson, 2011, p. 120)*

This proposed consensus views a CM as consisting of entities and activities organized in such a way that they are responsible for the phenomenon. When applied, explaining provides an intelligible answer to the question of why something happened and even more importantly how things work and how outcomes come about. All in all, the overall prospects for explanations based on CMs appear to be promising. In the following paragraphs we present the three elements and introduce their meaning: *responsible for the phenomenon, entities and activities,* and *organization.*

Firstly, the expression of being 'responsible for an phenomenon' in the above characterisation contains three aspects. The first aspect is individuation; 'responsible for an phenomenon' means that to explain a phenomenon we need to *individuate* the mechanism that causes it. A shared difficulty across the sciences is that a phenomenon may be caused by different CMs. The second aspect is diversity. 'Responsible for an phenomenon' refers to the diversity of things that CMs do, i.e., a single CM can cause different phenomena.. The meaning of 'diversity' refers for example to the behaviour of a standard roulette wheel. The wheel does not have different CMs for distributing the ball to pockets 16 and

17; instead the same CM produces the diversity of all 37 outcomes. The third aspect is that CMs carry out activities, such as regulation or control, exhibit behaviours, such as growth, and maintain stable states. Therefore, the expression 'responsible for an phenomenon' for causal inference means that one regards in CMs the importance, diversity and various forms of stability.

Secondly, a CM consists of entities and activities. Individuating a CM means individuating the entities and activities it is composed of. Each functional component in a CM has a *preferred* role in the production of a particular outcome. In this sense, the function of entities is tied to the *role* they play in the overall organization within the CM. A CM shows a combination of the components that jointly activate the CM, which, as a whole, produces the outcome or thing. Craver in [7] presented this statement in a simplified form, as shown in Figure 1a: where M stands for 'mechanism' and *Phi* for 'phenomenon. An entity that efficaciously engages in a productive activity acts as a cause (thus it is a difference maker) and a CM's activity to produce something (Φ) is explained by decomposing and analyzing how its components, that is, entities ($C_1, C_2, \ldots, C_n$) acting in a certain way ($\lambda_1, \lambda_2, \ldots, \lambda_n$), are relevant to Φ-ing. A CM should show that a component cannot be isolated from the other components; rather, its contribution to CM Φ-ing comes from its mode of operation, its size and force as well as its relation to the other components. The same functional component, for example, may have a different effect when it occurs in combination with other components.

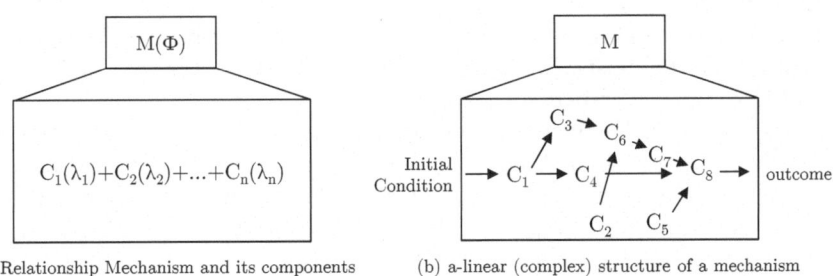

(a) Relationship Mechanism and its components   (b) a-linear (complex) structure of a mechanism

Fig. 1. Causal Mechanism adapted from Craver (2001)

Thirdly, in social research a CM producing a particular outcome rarely has a linear or stable structure, see Fig. 1b. The outcome is often produced through a complex system of functional components activating the CM to produce the outcome. Altogether, this complexity imposes operational conditions and restrictions for the identification and postulation of CMs. Operational conditions are important to detect since domain specific laws and entities explain nothing until initial conditions are specified: Newton's laws do not explain the movements of the planets unless their initial positions and velocities are specified. This may appear to be rather abstract but one needs to realise that employees of an organisation apply business rules (laws) in their activities implicitly or explicitly and we cannot predict how these activities will evolve unless we know something of their current state. For example the intake before medical surgery is prescribed in a protocol (e.g. check a persons identity, the surgery request and the actual medical status) by which we can tell what happens next to person X with age Y and request R. The concept of 'organization behind causality' is not only about *initial conditions* but also about *ongoing conditions* that allow the entities and activities to produce the phenomenon. 'Ongoing' is very important in this respect as initial conditions for laws (e.g. states and business rules) matter only at the beginning, while the organization behind causality matters throughout the operation of a CM.

The above discussion was aimed at showing how CMs can provide new insights for an account of explanation. By identifying the entities, the activities, and their organisation, we particularly focus on the *functioning* of the CM. It is the articulation of its functioning that provides the basis for a successful explanation. It is worth noting that CM-based explanation by and large assumes that what we describe is the functioning of 'worldly' CMs out there. The idea of decomposing the black box – figuring out the functioning of the CM – presupposes a mild form of realism causing the authors in the CM literature to believe i) that there is a box, and (ii) that it can be decomposed into smaller boxes. We follow the ideas of authors [6, 16, 18, 24] that CMs should occupy a central position in explanatory accounts. In Section 4, we elaborate on the relation between the decomposition mentioned here and the difference between functional and constructional perspectives as discussed in the Enterprise Engineering (EE) manifesto [12].

The idea to explain phenomenons by CMs can also be fruitfully applied to organizational diagnosis. In other words, a business process is composed of entities (departments or people) performing various types of activities. Their organisation causes the phenomenon we are interested it, be it the enterprise's typical behaviour or some kind of dysfunction. For example, the organization of a flower shop can be interpreted in mechanistic terms: different social individuals (entities) may work together and perform different roles (activities) in the management of the shop. Once a dysfunction in the behaviour of an organization is observed, we need to diagnose its origin or cause by identifying the point(s) at which the functioning of the mechanism clashes. Therefore, we need to identify: (i) the phenomenon itself; (ii) the entities and activities involved and (iii) a CM's organisation and operation (the role-functions of the entities producing the phenomenon) during diagnosis. In the next section, we present a case study that was performed using the current state of the art in organizational diagnosis. We will then analyze this case study to learn whether the current state of affairs in OD succeeds in explaining modelled using CM's.

3   Achieving a Mechanistic Explanation in OD

Thus far, we have focused on conceptions of a CM, without emphasizing how CMs are discovered in OD. The following case study covers an attempt to discover a CM. It illustrates that discovering a CM for a quality problem in a business process turns out to be far from a trivial problem in practice and presents the challenges of the current state of affairs of OD. The case study looks at the pension fund Zwitserleven which manages the pension arrangements of all its pension holders. In 2007 Zwitserleven decided to invest in a Lean Six Sigma (LSS) project to improve the *"Information Requests Handling"* (IRH) business process. The ambition of management was to reduce costs and increase customer satisfaction. From the CM perspective advocated in the previous section, we can reformulate this challenge as follows: "What is the CM that is responsible for the high costs and low customer satisfaction in IRH?". In the next sections we report our approach and we discuss the extent to which the performed analysis succeed in meeting the challenge in our formulation.

3.1   Identifying the Phenomenon to Explain

In this case, Zwitserleven applied the Lean Six Sigma (LSS) approach whose features are virtually all relevant tools and techniques that have been developed in industrial statistics [8]. Initially LSS requires a more specific description of the project goal by means of functional decomposition [8]. The CTQ-flowdown is the activity in which the CTQ-Tree is generated [9]. In such a tree, the quality focus (i.e., customer satisfaction) is specified by measurable quality variables which are Critical to Quality, see Fig. 2. In this case, customer satisfaction was specified by the quality variables: 'CTQ2: Throughput' and 'CTQ3: Rework'. The CTQ-tree template [read p. 47-p. 48 in   8] requires additional information on the quality variables, e.g. its unit, its measurement protocol, its null measurement, and its targeted value. This additional information is required to operationalise its measurement. The result of the CTQ-flowdown for Zwitserleven is presented in Fig. 2.

The CTQ-tree in Fig. 2 provides the necessary information to conduct the null measurement, required for an understanding of the current situation. The data of the null measurement consisted of 15295 information requests (IRs), the workload (CTQ1) between June 2006 and June 2007, which were subjected to statistical evaluation to better understand the IRH's behaviour. A classification of 13 types of IR's was used to conduct a pareto analysis. The analysis results showed that 6 out of 13 types identified created 80% of the workload (CTQ1), and furthermore, that the average throughput (CTQ2) was 6 days with a highly diverse distribution over all types. A histogram on CTQ2 showed that 29% of the IR were handled in one day, 20% in two days and 51% in three days or over. Rework (CTQ3) in this period (i.e., the percentage of IRs which were caused by a former IR) was 15%. A histogram on CTQ3 showed that 80% of this rework resided in four IR categories. Using this kind of analysis demarcates the problematic behaviour of business processes, but not the problematic phenomenon to be explained.

The availability of this analysis led to a management decision to improve IRH in one particular organisational department (CPA3). In the null measurement CPA3 was responsible for handling 7956 IRs. For the duration of four months the

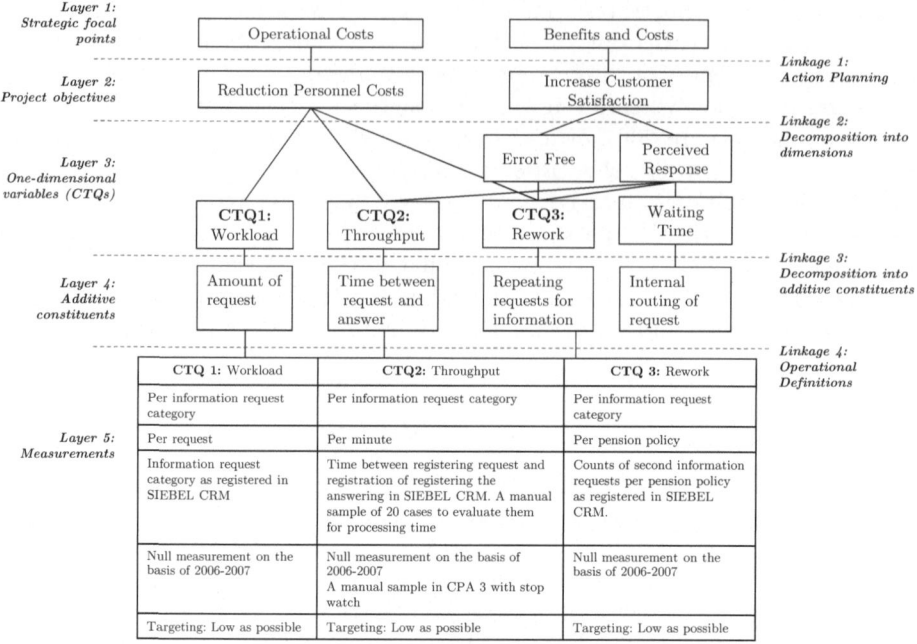

Fig. 2. CTQ Tree for Information Request Handling

project team was asked to assess and report monthly the quality level of CPA3. In the first month the improvement cycle was set up, followed by three one-month improvement cycles each. This period was considered to be sufficiently long to provide a reasonable conclusion whether the estimated savings could be achieved.

Nr.	Variable	Changes	Modelling Concept	Value Range
v1	**workload**	effect ctq1	[Customer Request]	(high..low)
v1.1	channel_type	change 1 .. 5	[Customer Request]	(post/fax/email/tel)
v1.2	file_type	change 1 .. 6	[Customer File, Policy]	(paper/digital)
v1.3	category	change 1 .. 3	[Information request]	(1..13)
v2	**throughput**	effect ctq2	[IRH Business Process]	(high..low)
v2.1	digital_format	change 1 .. 7	[Receipt and Registration]	(yes/no)
v2.2	clarity	change 1 .. 4	[Formal Letter]	(yes/no)
v3	**rework**	effect ctq3	[Completing Request]	(high..low)
v3.1	correct_policy	change 1 .. 4	[Policy]	(yes/no)
v3.2	correct_IR	change 1 .. 7	[Information Request]	(yes/no)

Fig. 3. Associations in IRH business process

The improvement cycle started with a brainstorm on the basis of a ishikawa/ fishbone diagram, a brainstorming technique asking: "which process variables influence the identified quality variables?" [8]. The brainstorm resulted in the identification of 30 process variables (e.g., 'availability of employees', 'new promotion activities', 'changes in regulations'). It was decided to observe 18 process variables, by registering their values (including the values of the quality variables) each time an IR was received. The correlation strengths between all 18 process variables and the 3 quality variables were determined by statistical software. The process variables with the most influence were filtered on the basis of their strength of correlation, and presented in a table, see Fig. 3. All entries of the table should be read as a tuple (e.g. <v1, v1.1>) of variables representing an *association* between a quality variable and a process variable. The reader should disregard the last two columns of this table since this information was added in the second phase.

The LSS team used the *associations* (e.g. <v1, v1.1>) as presented in Fig. 3 as input for suggestions to improve the values of the quality variables. The idea was to identify changes related to process variables (e.g. v1.1) in order to positively influence the CTQ (e.g. v1). During three months ideas were generated, implemented and statistically evaluated. The number of ideas is represented in the third column of Fig. 3. For example, one change was to start a pilot in which customers were called outside office hours (v2.2) to reduce the throughput time (v2). As a result more IR were handled within two days and customer satisfaction increased in November and December. However, the positive effect on V2 (CTQ2:Throughput) was temporary as calling pension holders outside office hours created high costs and inefficiencies, and the pilot, therefore, was terminated. To avoid such problems it was suggested to study the problem of IRH more fundamentally by reframing the identified association(s) as phenomenon(s) to be explained, and the next step was to identify the entities and activities involved in these phenomena modelled.

3.2 Identifying Entities and Activities Involved

CM theory (as discussed in Section 2) suggests that the next step to identify all entities and activities involved is the reframing of the identified associations (see Fig. 3). The assumption here is that the phenomenon is mechanistically produced. According to CM theory one should adopt the fallible, explanatory heuristics (as opposed to algorithms) of *decomposition* and *localization* to identify entities and activities [3]. in this theory decomposition refers to taking apart or disintegrating a CM into either component parts or component operations, and localization refers to mapping the component operations onto component parts. In the case study it was decided – based on previous experience – to apply the ArchiMate framework and its modelling language[1] for localization and decomposition.

---

[1] The authoritative source is the description by The Open Group and available on http://www.opengroup.org/archimate or as book see [19].

ArchiMate is based on the descriptive notion of architecture [15], which means that an enterprise architecture in ArchiMate corresponds to a functional model of the business processes in an enterprise. The ArchiMate framework [19, p.7] distinguishes three architectural layers: the business layer, the application layer, and the technology layer. Furthermore, the ArchiMate framework consists of a horizontal axis distinguishing three major aspects: the active structure, the behaviour, and the passive structure of a system. Each layer has its own meta-model: the business layer meta model, see [19, p.14], the application layer meta model, see [19, p.37], and the technology layer meta model, see [19, p.47]. Each meta model explains the core modelling concepts and the relationships between them. Each modelling concept belongs exclusively to one of the three aspects. As a result, the modeler chooses the system whose aspects prevail when representing a business process.

Decomposition with ArchiMate means looking for objects and artifacts of IRH that designate instances of the meta model concepts as presented in [19, p.14], positioning the identified entities within the ArchiMate framework and determining the relationships that exist according to the ArchiMate meta model. This was conducted in this case for both the business and the application layer. The result is a decomposition – a functional model – of IRH. Due to space limitations the ArchiMate model presented in Fig. 4 is a reduced version of the real model. We reduced its complexity to show that business objects were implemented in multiple data objects (e.g. the *business object* 'information request' is implemented in 15 *data objects* assigned to 6 *application components*). We, furthermore, did not present the application landscape in detail as the 42 information systems it consists of are too complex to present in this paper. Nevertheless, it must be taken into account since employees in IRH work with this application landscape.

Based on the functional decomposition of IRH, the ArchiMate model in Fig. 4, project team members conducted a localisation step. The exercise – as described in the theory of CM, see [3] – is one of mapping. Activities and entities are

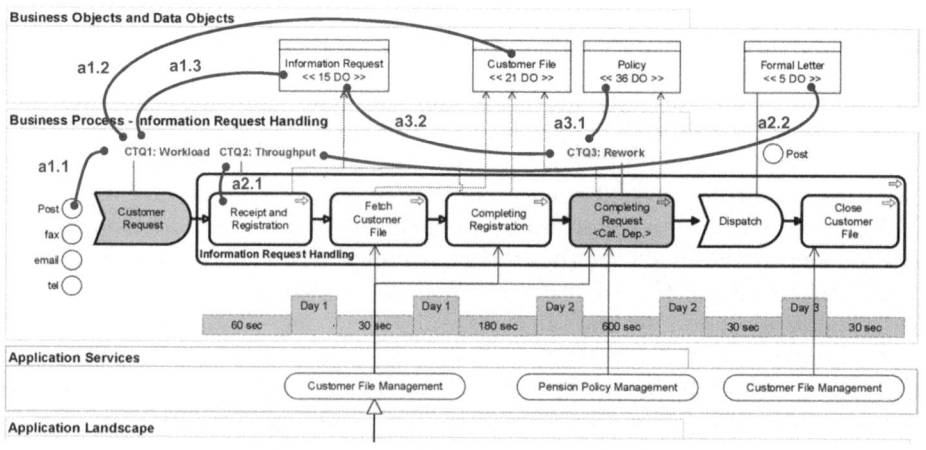

Fig. 4. ArchiMate model augmented with associations

paired on the basis of the relationship: "$<Variable>$ *is an aspect of* $<ArchiMate$ *Modelling Concept>*". The result of this localisation step is the table represented in Fig. 3. The table has, furthermore, been enhanced with information on the value range of the variables, which information was helpful for understanding the bandwidth of the behaviour studied. In the next phase the information captured in the table in Fig. 3 was used to augment the ArchiMate model. The result of this phase is presented in Fig. 4.

3.3   Identifying the Operation of the Mechanism

Since a CM is more than an aggregate of its parts, we must understand how entities and activities are organized to produce the phenomenon, or the behaviour of the system, read Sec. 2. The system that produces the behaviour of interest is decomposed into entities and activities, and it is important to distinguish the involvement of each part in the phenomenon in the next step. Inspired by Causal Loops Diagramming (CLD) [21] the project team decided to augment the Archi-Mate model with this type of diagram. The necessary information was already available from the former phase, see Fig. 3. The team expected to understand from augmentation how the involved entities and activities were organised and operated from this knowledge to understand how process variables influence quality variables. The strategy to identify the organisation and operation of CMs was that the involved entities and activities were 'traversed' by an association (e.g. ctq1 and a.1.1). The explanation was expected to reside in the ArchiMate model between the endpoints of the association, and the idea was that the modelling concepts and their relations would explain the operation of the CM.

In this case, the causal inference from correlation and a functional model was blocked. No plausible CM could be identified from the ArchiMate model that connected, for example, the rework of IR and its cause (e.g. $<v3, v3.2>$). It was not plausible to state that rework is explained with the notion of an interaction between the activities of 'completing registration' and 'completing request' (see Fig 4). Firstly, the relationship of 'one activity triggers the other' is a descriptive relationship, that does not provide us with any information on the operation of a business process. Secondly, ArchiMate does not recognise causality in its framework since it does not provide us with any theory for *interactions* between active elements and how these interactions cause *change*. One may argue that IRH was not modelled in active elements (e.g. actors and roles, see ArchiMate Framework [19, p.7]). We agree with this point. However, even if the IRH was modelled with active elements of the meta model of the business layer (read [19, p.14]) we would be confronted with *descriptive* relationships (e.g. a $<role>$ is assigned to $<actor>$), not with *causal* relationships. Such relationships do not represent interactions between active elements nor do they explain how state changes happen in IRH. A possible ArchiMate explanation for a CM connecting $<v3, v3.2>$ was therefore incompatible with what was known by the participants of the improvement brainstorm meetings. These participants introduced 7 changes for $<v3, v3.2>$ (see Fig. 3) mainly on the premise that social individuals (employees and pension holders) interact and they cause changes that influence the rework rate. A gap seemed to exist between the descriptive language and the causal knowledge that blocked causal inference from a model representing the business process under investigation.

4   Reflection

In this section, we reflect on the way the organizational diagnosis was performed in the case study using the theory of causal mechanisms. This reflection is based on the fact that, in OD, a diagnostician attempts to describe the functioning (and dysfunctioning) of an organization in causal terms. Such a causal description must be contrasted to any correlations identified. It should be noted, however, that the initial phases of the case study focused merely on identifying correlations (e.g., the correlations in Fig. 3). Though these correlations are useful to isolate the areas that are related to the phenomenon to be diagnosed, they are not sufficient to provide a causal description of the phenomenon, and what is needed, in fact, is an understanding of the organizational components that are to be changed to remedy a problematic phenomenon. As discussed in Section 2, the identification of a causal mechanism helps with this task. In the case study, an ArchiMate model was used precisely for this purpose. In this section, we will reflect on the ability of the ArchiMate approach to detect a CM.

In order to identify a mechanism, the underlying entities, activities and organization have to be uncovered, as discussed in Section 2. In an organizational diagnosis context, the adequate entities and activities related to the organizational phenomenon must be identified by means of a decomposition technique, after which we have to consider the organization of the entities and activities that produce the phenomenon. For this, two different perspectives can be used according to Enterprise Engineering: the functional perspective and the constructional perspective [12]. The functional perspective describes how the system is used by a certain stakeholder. Consequently, a functional model (or black-box model) used in this perspective is by its very nature subjective: the model can differ for each stakeholder. In other words, function is not a system property but a relationship between a system and a stakeholder. In contrast, the constructional perspective describes what a system *is*, in its ontological sense. A system is understood by its construction and operation, irrespective of how the system is used by stakeholders. A constructional model (or white-box model), therefore, can always be validated from the actual construction and thus its nature is objective.

To design a constructional model, EE requires a description of the composition, environment, boundary and activity of the system [11], based on the generic definition of a system [4]. The composition refers to the set of elements the system consists of (i.e., the elements required in the CM definition). The activity refers to the state changes caused by the system (i.e., the activities required in the CM definition). The structure refers to the way the elements of the system influence each other (i.e., the organization required in the CM definition). Consequently, only a constructional (and not a functional) perspective is adequate to uncover the relevant entities, activities and their organization for describing a CM. Based on the case study in this paper, we argue that such a perspective is currently lacking in organizational diagnosis. Three observations in the case study support this argument.

A first observation is that only *functional decomposition* is typically used. On a black-box model, functional decomposition can be applied to allow a focus on a sub-part of the system. It is important to realize that a functional decomposition only results in other models from a functional perspective. A fine-grained functional decomposition is not the same as creating a constructional model, since this perspective requires the elements described in the previous paragraph (i.e., composition, environment, boundary and activity). Therefore, functional decomposition does not allow us to identify causal mechanisms. Indeed, EE argues that a functional perspective is sufficient to control the behaviour of a system, but not to change the system itself [10]. Similarly, CM authors describe how Causal Loops Diagramming (CLD) can be used to specify associations between variables [20], but that such approaches are only able to understand the behaviour of organizations, not to explain the observed phenomenon [6]. Moreover, Woodward argues that such approaches may even fall short when used for predicting behaviour: *"without constructional knowledge, it is not possible to foresee the conditions under which those relations might change or fail to hold altogether"* [27]. The case study shows that current OD approaches rely heavily on functional decomposition. As an example, we mention the CTQ-tree, which decomposes goals into finer-grained elements, but does not attempt to define the organizational components needed to fulfil the goals. These approaches are insufficient to identify a causal mechanism.

A second observation is that no clear modelling concepts (i.e., entities and activities) are used when elaborating on functional decomposition models such as the CTQ-tree. In the case study, an ArchiMate model was used for this purpose. While an explicit meta-model is presented in ArchiMate (e.g. the business layer meta model [19, p.14]), it is clear in the case study, that this tends to 'blur' reality with 'fictional' modelling concepts. Many ArchiMate modelling concepts and relationships are conceptual and not observable (see [19, p.14]), do not correspond to real entities and activities or are too abstract to be useful for causal inference. For example, modelling a "Customer File Management" as an [application service] in Fig. 4 makes it very difficult to establish a direct relationship with observations, as such a service does not exist in reality. Instead, a large number of applications perform this service. Other examples are relationships such as "triggers", "realises" and "assigned to" which are used to indicate a kind of activity without providing any details on how that activity is carried out. Additional problems occur when the provisional status of the applied modelling concepts and relationships disappear. Both examples show to what extent a functional model can be detached from reality, and can lead to an incorrect diagnosis. Adding such modelling concepts can easily result in "an illusion of understanding" [6]. What is required instead is a meta-model that describes the adequate entities and activities for the phenomenon under diagnosis. This meta-model is adequate if it relies on an (inter)action-theory with activities caused by entities in a business process as subject matter.

A third observation is that the use of inadequate modelling concepts idealizes reality, instead of adequately describing it. This obstructs the correct identification of a CM. By modelling a business process in the ArchiMate language one implicitly assumes that a business process is a sequence of activities. This presupposition is embedded in ArchiMate's meta model (see the business layer meta model in [19, p.14]), and has to be accepted implicitly by any modeller. Another presupposition of ArchiMate is that every employee is allocated to some activity. Both presuppositions force the diagnostician to see a business process as a properly organized system, while in reality, employees communicate and conduct activities outside the organization as well. Again, this shows that diagnosticians require a meta-model both to adequately describe the system producing the phenomenon and to empirically validate said phenomenon. If not, what is actually modelled is an idealized reality omitting important causal information. From the analysis of the organization of the system, therefore, it must be evident which entities and activities are to be included or excluded in a certain meta-model, as, otherwise, a diagnostician cannot judge whether a model allows the detection of the CM for a certain phenomenon.

Based on these observations, we argue that a constructional perspective with a clearly defined meta-model must be integrated in organizational diagnosis approaches. Approaches such as EE, which explicitly incorporate a constructive perspective and separate it from behavioural observations, do not offer any practical support on how to use both perspectives to find a CM. However, the hypothetico-deductive methodology of causal models does provide a way to integrate both perspectives. It involves three stages: (1) hypothesising, (2) building the model, and (3) drawing conclusions on the empirical validity or invalidity of the model [20]. The fact that in hypothetico-deductive methodology, behavioral measurements belong to a "factual world", while the constructional model belongs to an "interpretative world" is important for this reflection. Similarly, EE argues that all behaviour is engendered by the construction [10]. Therefore, the behaviour of a system as a CM can only be understood through an alternation between the functional and the constructional perspective. Based on our reflection and supported by theories on CM and EE, we conclude that alternating between function and construction is crucial for identifying the CM responsible for a certain phenomenon. In future research we will, therefore, focus on the construction of a method which enables such an alternation and adheres to EE in order to detect a CM.

5   Contributions and Conclusion

Organizational diagnosis is a subfield in EE interested in finding effective procedures for both the identification of dysfunctions in organizations and interventions to improve organization performance. While the common practice of OD presents some standard methods, for instance LSS, we argue that these can be greatly improved if integrated with concepts coming from philosophies on causality. In particular, CMs can enhance diagnostic procedure as they provide important information on the structure and functioning of an organization. We proceeded to present the theories on CMs in order to establish a common understanding to be applied in an exploratory case study. In this exploratory case study we enhanced LSS with the functional modelling approach of ArchiMate, which is required in order to move towards adopting the explanatory power of a CM. It was shown that functional modelling – on the basis of ArchiMate – is not ideal to detect a CM. We conclude that an ontological perspective on business processes is needed.

With this paper, we hope to open new paths for the professionals in OD. We recognise that the integration of concepts coming from different fields may be difficult and will take time, but we believe that it will be a beneficial exercise. Indeed, such exercise will be beneficial to both communities. On the one hand, the use of CMs, as said before, can improve the intelligibility of the explanations from OD. On the other hand, the 'causality in the sciences' literature has not investigated the field of EE, so fruitful exchanges can be foreseen in this direction as well. Nevertheless, one should note that within the school of EE a theory to capture the ontology of a business process does exist. CM can be revealed from such an ontology, however, it was never used for diagnosis since EE is focussed on designing and engineering business processes. Furthermore, the theory on the ontology for business processes is operationalised in the DEMO^2 approach, which includes a modeling language. Effort should be invested to study the use of DEMO in a meaningful, evidence-based way for OD. Most importantly, OD professionals must begin the process of organizing and sharing what they know to inform and expand the knowledge that will move OD towards an approach aiming for explaining by a CM.

To the extent that adopting CM as an explanatory power that goes beyond traditional statistical evidence, a CM driven OD approach represents a significant change in management science and practice. We propose to achieve that change: both philosophers on causality and scientists of management science will hopefully realize that current diagnostic practices in management science are not working (discrepancy); that evidence-based diagnosis on the basis of statistics is the correct path (appropriateness); that intelligible explanation requires an ontological perspective (efficacy); that leaders in both fields are committed to change (principal support); and that change is beneficial to themselves (valence). The authors hope to encourage a change in practice by acknowledging the issues of discrepancy that have emerged in OD literature and by shedding some light on both the appropriateness and efficacy of explaining by CMs.

References

[1] Alderfer, C.P.: The Methodology of Organizational Diagnosis. Professional Psychology 11(3), 459–468 (1980)
[2] Alderfer, C.P.: The Practice of Organizational Diagnosis: Theory and Methods, 1st edn. Oxford University Press, USA (2010)
[3] Bechtel, W., Richardson, R.C.: Discovering Complexity. Princeton University Press (1993)
[4] Bunge, M.A.: Treatise on Basic Philosophy. Ontology II: A World of Systems, vol. 4. Reidel, Boston (1979)

---

^{2} Design and Engineering Methodology for Organisations.

[5] Bunge, M.A.: How Does It Work?: The Search for Explanatory Mechanisms. Philosophy of the Social Sciences 34(2), 182–210 (2004)

[6] Craver, C.F.: When mechanistic models explain. Synthese 153(3), 355–376 (2006)

[7] Craver, C.F.: Role Functions, Mechanisms, and Hierarchy. Philosophy of Science 68(1), 53–74 (2001)

[8] de Koning, H.: Scientific Grounding of Lean Six Sigma's Methodology. PhD thesis, UVA Amsterdam (2007)

[9] de Koning, H., de Mast, J.: The CTQ flowdown as a conceptual model of project objectives. Quality Management Journal 14(2), 19 (2007)

[10] Dietz, J.L.G., Hoogervorst, J.A.P.: The Principles of Enterprise Engineering. In: Albani, A., Aveiro, D., Barjis, J. (eds.) EEWC 2012. LNBIP, vol. 110, pp. 15–30. Springer, Heidelberg (2012)

[11] Dietz, J.L.G., Mulder, H.B.F.: Organizational transformation requires constructional knowledge of business systems. In: HICSS 1998: Proceedings of the Thirty-First Annual Hawaii International Conference on System Sciences, vol. 5, p. 365. IEEE Computer Society, Washington, DC (1998)

[12] Dietz, J.L.G. (red.): Enterprise Engineering The Manifesto (2011), http://www.ciaonetwork.org/publications/EEManifesto.pdf

[13] Harrison, M., Shirom, A.: Organizational diagnosis and assessment: Bridging theory and practice. Sage (1998)

[14] Hempel, C.G., Oppenheim, P.: Studies in the logic of explanation. In: Hempel, C.G. (ed.) Aspects of Scientific Explanation and Other Essays, pp. 245–282. Free Press, New York (1965)

[15] Hoogervorst, J.A.P., Dietz, J.L.G.: Enterprise Architecture in Enterprise Engineering. Information Systems Journal 3(1), 3–13 (2008)

[16] Illari, P.M., Williamson, J.: What is a mechanism? Thinking about mechanisms across the sciences. European Journal for Philosophy of Science 2(1), 119–135 (2011)

[17] Machamer, P., Darden, L., Craver, C.F.: Thinking about mechanisms. Philosophy of Science 67(1), 1–25 (2000)

[18] Mouchart, M., Russo, F.: Causal explanation: recursive decompositions and mechanisms. In: Illari, P.M., Russo, F., Williamson, J. (eds.) Causality in the Sciences, pp. 317–337. Oxford University Press (2011)

[19] The Opengroup. ArchiMate 2.0 Specification. The Open Group (2012)

[20] Russo, F.: Causality and Causal Modelling in the Social Sciences: Measuring Variations. Methodos Series. Springer (2009)

[21] Russo, F.: Correlational Data, Causal Hypotheses, and Validity. Journal for General Philosophy of Science 42(1), 85–107 (2011)

[22] Salmon, W.C.: Four Decades of Scientific Explanation, vol. 3. University of Minnesota Press (1989)

[23] Salmon, W.C.: The Importance of Scientific Understanding. In: Causality and Explanation, pp. 1–17. Oxford University Press (January 1998)

[24] Sankey, H.: Scientific Realism: An Elaboration and a Defence. Theoria A Journal of Social and Political Theory 98, 35–54 (2001)

[25] Struss, P.: Fundamentals of model-based diagnosis of dynamic systems. In: Proceedings of the 15th International Joint Conference on Artificial Intelligence, vol. 15, pp. 480–485. Lawrence Erlbaum Associates ltd. (1997)

[26] Wagner, C.: Problem solving and diagnosis. Omega 21(6), 645–656 (1993)

[27] Woodward, J.: Making Things Happen: A Theory of Causal Explanation. Oxford University Press, USA (2005)

Transformation of Multi-level Systems – Theoretical Grounding and Consequences for Enterprise Architecture Management

Ralf Abraham[1], José Tribolet[2,3], and Robert Winter[1]

[1] University of St. Gallen, Institute of Information Management,
Mueller-Friedberg-Strasse 8, 9000 St. Gallen, Switzerland
[2] CODE, Center for Organizational Design & Engineering, INOV, Rua Alves Redol 9,
Lisbon, Portugal
[3] Department of Information Systems and Computer Science, Instituto Superior Técnico,
Technical University of Lisbon, Portugal
{Ralf.Abraham,Robert.Winter}@unisg.ch,
Jose.Tribolet@inesc.pt

Abstract. In this paper, we investigate the support of enterprise architecture management (EAM) for enterprise transformation. Conceptualizing enterprises as systems, we draw on two theories that investigate static and dynamic system aspects, respectively – the theory of hierarchical, multi-level systems and control theory. From the theory of hierarchical, multi-level systems, we first introduce three orthogonal dimensions of hierarchy – layers, strata, and echelons. We then position EAM as a cross-dimensional transformation support function in this there-dimensional hierarchy space. Finally, we draw on control theory to derive a model of control and feedback loops that enables a designed EAM support of system-wide transformations. Using this model, we propose to extend the multi-level systems theory by a set of interlinked feedback loops as a fourth dimension. A case study of transformation in the Portuguese air force serves as an example illustrating the usefulness of the two theories for describing enterprise transformation.

Keywords: Enterprise Architecture Management, Control, Feedback, Multi-Level Systems, Hierarchy.

1   Introduction

Increasing variety in their environment forces enterprises to change themselves at an ever higher pace. Sources of variety in an enterprise's environment include economic pressures from competitors, as well as politically, socially or technologically-induced changes. We understand enterprise transformation as designed and fundamental change, in contrast to ad-hoc, routine change. Enterprise transformation is a purposeful steering intervention into an enterprise's evolution, in order to respond to perceived opportunities, deficiencies or threats [1]. Despite the relevance of enterprise transformation, a big number of these transformation efforts fail. Reports indicate

H.A. Proper, D. Aveiro, and K. Gaaloul (Eds.): EEWC 2013, LNBIP 146, pp. 73–87, 2013.
© Springer-Verlag Berlin Heidelberg 2013

failure rates ranging from 70 to 90 per cent, across a broad range of domains [2]. These failures are often traced back to mistakes in strategy implementation and the coordination of the actual transformation efforts. In order to successfully implement enterprise transformation, Dietz and Hoogervorst [2] argue that a constructional, white-box understanding of enterprises is required in addition to a functional, black-box understanding.

One approach that is concerned with an understanding of enterprise construction is enterprise architecture (EA). The purposeful design and change of EA according to strategic goals is the concern of enterprise architecture management (EAM). By defining principles to restrict design freedom (and thereby guiding design), one of the core tasks of EAM is to coordinate enterprises transformation [3, 4, 5]. Since EA focuses both on results (e.g., models) and activities (e.g., principles) in designing an enterprise, we draw on two theories that focus on static and dynamic system aspects: The theory of hierarchical, multi-level systems and control theory. When applied to enterprises, the former theory is concerned with enterprise construction, and the latter with enterprise transformation.

Within his theory of hierarchical, multi-level systems, Mesarovic [6] distinguishes between three orthogonal notions of hierarchy: Strata, layers, and echelons. By following this explicit distinction, we are able to define the positioning of EAM in an enterprise by analysing it from multiple hierarchy angles. We aim to show that in order to support enterprise transformation, EAM must employ a more differentiated understanding of hierarchy. To break down the resulting three-dimensional hierarchy space (strata, layers, and echelons) to the specific purpose of describing enterprises, we will provide three exemplary organizational design and engineering (ODE) approaches that each focus on one dimension of hierarchy in particular. We then position EAM in this framework and show how it cuts across these hierarchical dimensions.

Having an understanding of the static aspects of enterprises and the positioning of EAM, we describe an enterprise from a dynamic perspective as comprising three different kinds of feedback loops. We identify enterprise transformation as a special instance of a feedback loop, as purposeful and designed change in contrast to both permanent, evolutionary adaptation and sudden improvisation. The research questions we address are the following:

1. How can a conceptualization of enterprises as hierarchical, multi-level systems improve the effect of EAM for their transformation?
2. How can transformation be grounded on multi-level systems theory and control theory?

The rest of this paper is organized as follows. Section 2 introduces the theory of hierarchical, multi-level systems, basics of EAM and provides a framework of hierarchy and a positioning of EAM. Section 3 discusses the concept of feedback loops. In section 4, a case study of the Portuguese air force illustrates both theories applied to a real organization. Section 5 discusses implications for EAM and offers a conceptualization of feedback loops as another dimension of hierarchy in Mesarovic's [6] terms.

Section 6 summarizes related work, before section 7 discusses limitations and provides a conclusion.

2   Framework of Hierarchy

2.1   Theory of Hierarchical, Multi-level Systems

To establish a framework for describing enterprises, we must first distinguish two contexts of hierarchy: In a management context, hierarchy means a relationship of authority and responsibility between higher and lower level units. This implies that higher level units have authority to delegate tasks to lower level units (which are required to carry out these tasks), but they must at the same time bear responsibility for their actions. Therefore, the concepts of authority and responsibility are inextricably linked. By contrast, in an engineering context, hierarchy refers to a vertical decomposition of a system into subsystems (i.e., a vertical arrangement of subsystems).

Hierarchy is primarily used in the latter context by Mesarovic [6] in his theory of hierarchical, multi-level systems. Mesarovic distinguishes three orthogonal notions of hierarchy to describe a system: Strata, layers, and echelons. Strata and layers refer to a vertical decomposition of the system (i.e., a decomposition of the overall system into subsystems), while echelons are a horizontal decomposition to coordinate and integrate activities of various decision units. To refer to any of these notions, the generic term 'level' is used. Fig. 1 is adapted from Mesarovic [6] and shows the three different notions of hierarchy combined to describe an enterprise. Note that both strata and layers can be used to decompose the entire system, but also to decompose the subsystem that is under the control of a certain decision unit.

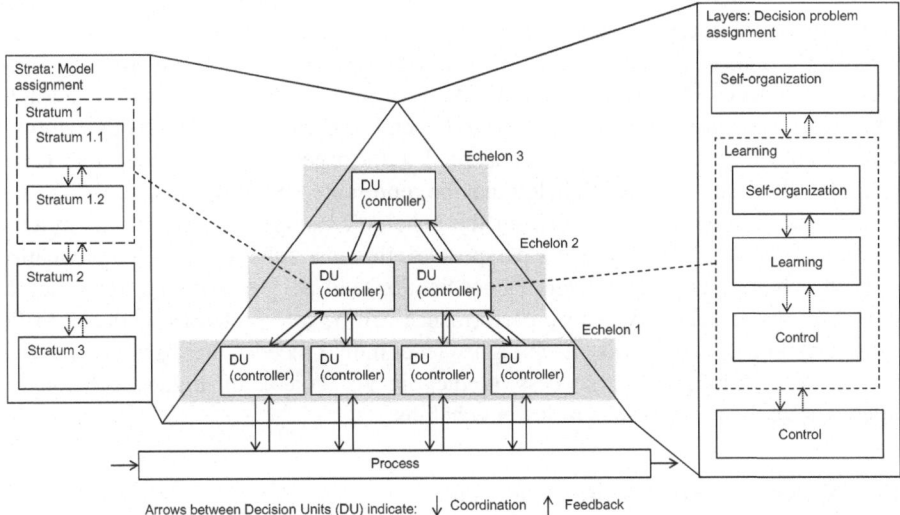

Fig. 1. Relationships of hierarchy notions in multi-level system (adapted from Mesarovic [6])

Despite the differences between the three notions of levels, there are some important commonalities: (1) Higher-level units are concerned with larger chunks of the overall system than lower-level units. (2) Higher-level units are concerned with the slower aspects of system behaviour; i.e. they are taking fewer decisions and intervene into system behaviour less frequently. (3) Problem descriptions on higher levels contain more uncertainty and are harder to formalize quantitatively than problem descriptions on lower levels.

The first notion of hierarchy, strata, describes levels of problem description or abstraction. While a model on a higher stratum provides a better understanding of overall system behaviour, a model on a lower stratum provides a more detailed functional explanation of individual system parts. What is considered as a whole system on a lower stratum may form a subsystem on a higher stratum. Strata thus describe the same entity on different levels of granularity. They may use a different form of modelling on each level, to capture the model contents in a suitable way.

The second notion of hierarchy, layers, describes levels of decision complexity. Since decision problems on lower layers can be expressed by more formal methods of description in the strata dimension, more formal techniques can be used to solve these problems. For example, in a manufacturing plant, a decision problem concerning optimum production sequences may be addressed by numerical optimization techniques (search layer), whereas decision problem on higher layers, such as which products to offer and how to market them (self-organizing layer), rely on less formal techniques such as heuristics. Layers reflect a hierarchy of goals, where the solution of goals on a higher layer decreases uncertainty on underlying layers. Put another way, lower layers need higher layers to limit their solution space – decisions made on higher layers limit the degree of design freedom on lower layers.

The third notion of hierarchy, echelons, is specific to organizations. Mesarovic [6] refers to echelons as a horizontal decomposition and calls the resulting system an organizational hierarchy. The term "organizational hierarchy" implies the use of hierarchy in the management context. On each echelon, there are one or more decision units. These decision units may be individuals or groups of individuals (e.g., an individual acting as project manager or an architecture board). As with the other notions of hierarchy, higher echelons are concerned with larger system aspects than lower echelons. The task of a given echelon may be represented by a stratified description of the subsystem under its control (from the echelon's perspective, this is the system, whereas from the overall system's perspective, this is a subsystem). Communication between echelons takes the form of coordination and performance; higher echelons coordinate lower echelons, whose performance can be interpreted as feedback to the higher echelons. Coordination from higher-level units precedes feedback from lower-level units. Consequently, success (achieving its goals) of higher echelons is dependent on the performance of lower echelons

2.2   Enterprise Architecture Management

According to the ISO/IEC/IEEE Standard 42010, architecture is defined as "the fundamental organization of a system, embodied in its components, their relationships to each other and the environment, and the principles governing its design and evolution" [7]. This definition of architecture involves two aspects: The first part of the definition forms a descriptive aspect, concerning the structure of the system's building blocks and the relationships between them. The second part ("[...] the principles [...]") forms a prescriptive aspect, effectively restricting the design and evolution space of the system under consideration. EAM is a continuous management process concerned with establishing, maintaining and purposefully developing an enterprise's architecture [8, 9]. From the IEEE definition, architecture is concerned both with results (descriptive aspect) as well as with guiding activities leading to these results (prescriptive aspect). In terms of artifacts, EAM provides models on the current state of an enterprise (as-is), the future state (to-be), as well as a transition plan on how to get from the current to the future state.

Addressing the descriptive aspect of architecture, EAM is concerned with establishing transparency. Capturing the current state of EA and keeping this information up-to-date is therefore seen as one of the EAM team's core tasks [8, 10]. Concerning the prescriptive aspect of architecture, EAM is concerned with maintaining consistency. Principles guide enterprise evolution by restricting design freedom [2] in order to maintain consistency between the enterprise strategy and its implementation (i.e., the actual EA). The management function of EAM is a good example to illustrate the interplay between strata, layers, and echelons:

By focusing on a high level of abstraction (high stratum) in the descriptive aspect, EAM is able to provide a holistic overview of the enterprise. In order to understand individual aspects of an enterprise in greater detail (e.g., the technical infrastructure), partial architectures have to be relied upon. This approach is referred to as "broad instead of deep": EAM cuts across several decision layers in an enterprise and provides suitable models of each layers' concerns on a high stratum. Examples of decision layers may be found in the business engineering framework [11], ranging from strategy, organisation, alignment to software and infrastructure layers.

In the prescriptive aspect, EAM is concerned with restricting design freedom by providing architectural principles. EAM principles support a layer notion of hierarchy as described by Mesarovic: Decisions made on one layer restrict the search and solution space of lower layers. For example, a principle advocating the use of commercial off-the-shelf software on the organisation layer narrows down the solution space on the software layer by removing the option of in-house development (and the need to select a particular programming language of platform).

Finally, echelons describe governance hierarchies – which organizational entities have the right to make which decisions, and thus are able to influence the actions of organizational entities on a lower echelon. The cascade of architectural principles can therefore be seen as a representation of the organizational governance hierarchy: Who has the right to govern whose decisions – which principle owners (organizational units responsible for the formulation, justification and maintenance of a given principle) may restrict the design freedom of other organizational units.

2.3   Framework of Hierarchy and Positioning of EAM

To create a framework for EAM positioning, we first consider three ODE approaches that each illustrate one of the hierarchy notions previously discussed.

For the strata dimension, consider the architecture landscapes in the TOGAF framework [12, p. 481]. TOGAF contains architectural views on the enterprise with varying levels of granularity, namely strategic architecture describing a long-term, highly aggregated view on the enterprise, segment architecture focusing on a more detailed description of areas within the enterprise, and finally capability architecture to describe operational competencies.

We illustrate the layer dimension using the aspect organizations found in DEMO [13]. Enterprises are regarded on three different aspect organizations, namely the business organization (B-Organization), the intellect organization (I-Organization) and the document organization (D-Organization). These organizations form a hierarchy of decision layers, with higher layers setting a frame for lower layers and services from lower layers supporting the operation of higher layers: A redesign of the B-Organization results in changes to the I- and D-Organization. Bottom-up, the D-organization supports the I-Organization, which in turn supports the B-Organization.

The echelon dimension is exemplified by the total information systems management (TISM) approach [14]. Management of information systems is broken down to five levels. Strategic guidelines, IS framework, IS Project Portfolio, IS Project and IS support. We will leave out IS support since this level focuses on local user support only instead of enterprise-wide aspects.

Each of these echelons is represented by organizational actors carrying out assigned roles with authority and responsibility. Organisational actors may either be individuals in roles or groups of individuals acting as boards. We shall call the area of authority and responsibility of organizational actors their domain. For example, the management board defines strategic guidelines, an architecture board defines the IS framework (in Österle et al.'s [14] definition, an IS framework covers not only data and functions of electronic information processing, but also the organizational dimension. This is in line with the notion of EAM extending beyond IT to also include business aspects), a project portfolio management board sequences individual projects, and finally project management teams carry out individual projects that generate new or improve existing capabilities.

We position EAM as shown in Fig. 2 on the highest stratum, the strategic architecture level. Lower strata are covered by detailed architectures such as segment or capability architectures. This is consistent with the idea that models on higher strata provide an explanation of the overall system behaviour, while lower-level models such as segment or capability architectures provide a more detailed functional explanation of subsystems.

While all organizational actors take part in EA (by creating organizational reality within their domain, where they enjoy freedom of action), the purposeful evolution of EA (i.e., its management: EAM) is the main task of a specific echelon, the EAM board. By using architectural principles to limit design freedom, EAM is a cross-layer approach. When principles are operationalized as concrete standards, they are used to guide enterprise design across all layers, ensuring that lower layers support higher layers.

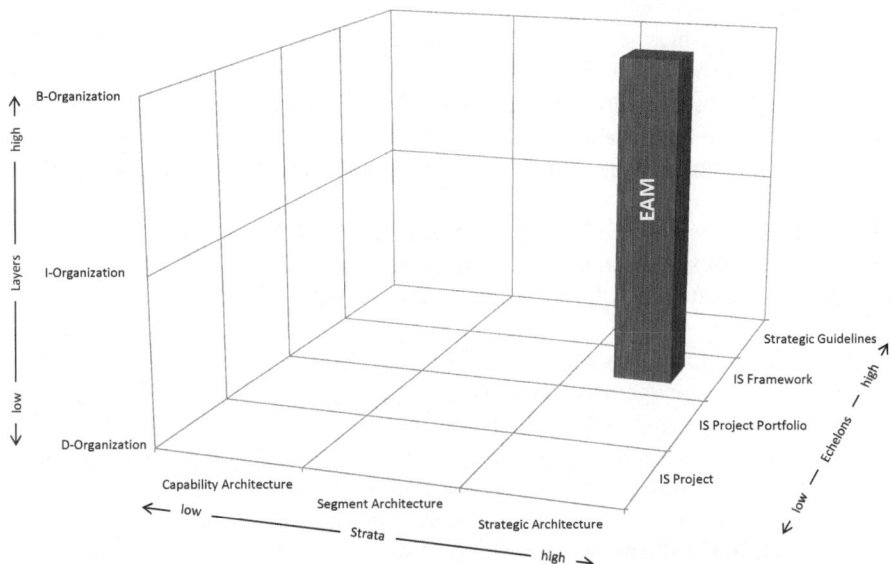

Fig. 2. Positioning of EAM in three-dimensional hierarchy space

3   Feedback Loops

In order to describe dynamic aspects of a system, we turn to control theory and feedback loops [15]. Control is defined by Åström and Murray as "the use of algorithms and feedback in engineered systems" [15]. Feedback is a key component in improving a system's robustness against uncertainty. A simple example would be a feedback system to control the speed of a car, e.g. when cruise control is turned on. The actual speed of the car is observed by a sensor, and if deviations are sensed, the flow of petrol to the engine is regulated. Fig. 3 illustrates this example of a feedback loop with an observer (Sense Speed), a modeller (Compute) and a controller (Actuate Throttle).

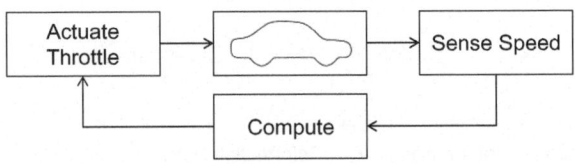

Fig. 3. Example of a feedback loop from Åström and Murray [15, p. 18]

By constantly observing system states and comparing observed data against system goals, a controller component can compute corrective measures and then change system variables to re-align to system goals. A feedback loop consists of an observer that records environmental data, a modeller that interprets the data and calculates corrective actions, and a controller that influences the system based on the input from the modeller. Thus, the actual car speed can be kept as close to the desired speed as possible. The example also shows that a system being observable is necessary for the system to be controllable: If the car speed cannot be sensed, it cannot be controlled. However, being observable is not sufficient for being controllable: If only the sensor worked, but not the components that compute speed adjustments and actuate the throttle, the car would still be uncontrollable. Furthermore, not all observable variables are also controllable.

In a complex system, there may be several feedback loops operating in parallel: Next to the feedback loop concerned with the car's speed, there are also several other loops, e.g. for climate control, regulating the air condition in the passenger cabin.

Like a travelling car, an enterprise can also be considered a system in which several feedback loops run in parallel. This theme of control is also central to one of the understandings of management: That as the structuring, control and development of productive social systems such as enterprises [14, p. 22]. Note that there are also other understandings of management, such as the behavioural notion that focuses on getting activities done by people. However, for the purpose of this paper, we will follow the notion of management as a cyclic feedback loop that comprises the activities plan, do, check, and act. Fig. 4 illustrates a hierarchical structure of three feedback loops.

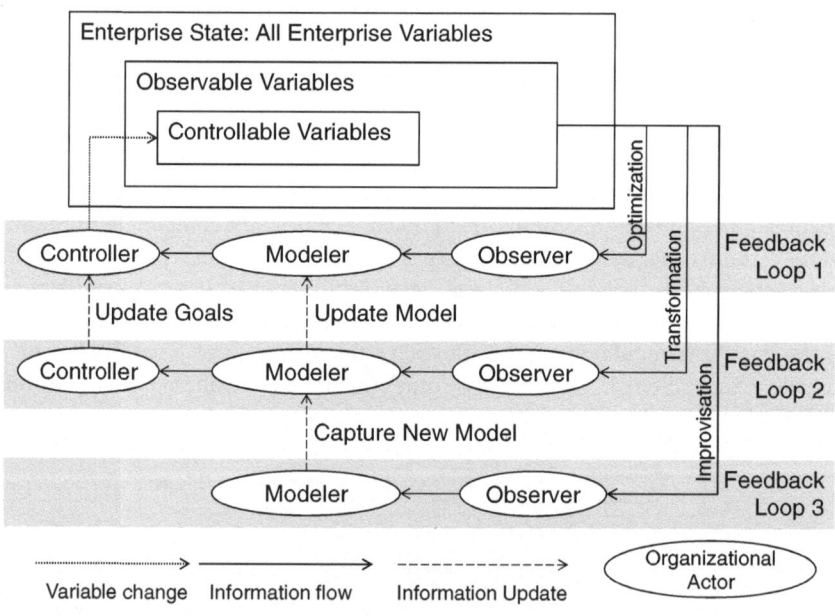

Fig. 4. Feedback loops in enterprises

Management in general, and also management of EA, is mandated by the fact that enterprises are never in stasis, but are constantly subjected to environmental turbulences. Three dimensions of environmental turbulence are suggested [16, 17]: Frequency, Amplitude, and Predictability. We consider an enterprise as having a certain state at each moment in time, and consisting of a set of observable and a set of controllable variables. Depending on the degree of environmental turbulence the enterprise encounters, one or several of the following feedback loops are triggered.

Feedback loop 1 is the basic loop that runs continuously. It deals with on-going optimization, with running the business. Turbulences occur frequently, but with low amplitude, and they are predictable. Based on existing models (e.g., architectural and process models), deviations are observed by organizational actors and control measures are initiated. These models describe the enterprise on different strata. It is important to note that all organizational actors run this first feedback loop within their own domains. Taking the example of a car company, inventory management or production sequencing on the assembly floor would be exemplary activities in feedback loop 1, as would be running IT systems in the administrative departments. In feedback loop 1, enterprises react to expected exceptions. This is also discussed as resilience by Aveiro et al. [18].

Feedback loop 2 is triggered when unexpected exceptions or environmental turbulences cause a change in models. In this case, turbulences occur less frequently than in feedback loop 1, but they have higher amplitude. Still, they are predictable and thus allow for initiating a planned, purposeful transformation. Therefore, enterprise transformation takes place in this loop. Enterprises continue running their business (i.e., conducting daily operations within feedback loop 1), but they take additional measures to transform themselves: They enter feedback loop 2, formulate transformation goals and initiate a series of projects that will eventually change the enterprise state (Update Model / Update Goals). When transformation is complete, enterprises exit feedback loop 2 and continue running feedback loop 1 which now operates on a new enterprise. That is, at the end of feedback loop 2, enterprises have reconfigured their operational resources to achieve a fit with the new environment [19]. They have also updated their models to reflect this new environment. The development of a new generation, fuel-saving engine would be a concrete example in a car company.

Feedback loop 3 is triggered when an enterprise is faced with environmental turbulences that require immediate reactions. These turbulences are characterized by high amplitude and virtually no predictability [20]. As opposed to feedback loop 2, environmental change in this feedback loop is unpredictable and therefore requires quick, improvisational actions that result in the capture of a new model [21]. Once an enterprise has made sense of the new situation (i.e., models and goals have been updated), it goes back to feedback loops 2 and 1 in order to initiate the necessary transformation projects and continue running its business.

After feedback loops 2 or 3 have handled unknown exceptions for the first time, these exceptions are no longer unknown and can in the future be dealt with by a lower feedback loop, e.g. by feedback loop 1. In other words, enterprises that have experienced certain environmental turbulences may learn from these experiences if they are able to add them to their pool of known exceptions [18, 19]. If enterprise goals were never questioned, and enterprise models never updated, enterprises would be limited to single-loop learning. A hierarchy of feedback loops allows for updating enterprise goals and models and enables double-loop learning in organizations [22]: modifying variables based on previous experiences.

The following example may illustrate the transitions between different feedback loops and the potential for organizational learning: Consider a car company is hit by a strike from one of its suppliers. It may try to respond to this situation within feedback loop 1, by attempting to secure the required material from another source. If this fails, it may enter feedback loop 2, starting negotiations with unions to reach a settlement. If this also fails, it may enter feedback loop 3 and try to change its entire logistics from just-in-time supply to in-stock supply. In this case, the company may no longer be threatened by further exceptions of the same kind (strikes) – this kind of exception is from now on included in the pool of known exceptions and can in the future be dealt with by feedback loop 1.

4   Case Study: Transformation in the Portuguese Air Force

To give a practical example of feedback loops and different notions of hierarchy describing an enterprise transformation, we consider the example of the Portuguese air force case study as reported by Páscoa and Tribolet [23]. This organization operates different weapons systems that can be characterized by their degree of technological advancement. In normal operating mode, the mission of the air force includes the defence of the national airspace, air transport operations, as well as search and rescue missions. Its core business process is flying. Being a non-profit organization dependent on state funding, the Portuguese air force is developing a strategy map to maximize its core output – the number of flying hours – given a fixed annual budget. The overall goal is broken down into individual objectives like improving drafting of personnel, or providing more efficient aircraft maintenance processes.

Considering the strata notion of hierarchy, the Portuguese air force has developed a number of models like an objectives model (providing a mapping of objectives to business models), descriptions of business processes (e.g. flight operation, personnel training, and public relations), or performance indicators and dashboards providing information on various organizational subsystems (e.g., describing the number of people involved with a specific weapon system or the number of missions performed by given air craft or squadrons). Using this stratified description, several scenarios (organizational configurations) can be described to identify the impact of possible budget cuts.

On the layer dimension, a set of business rules and policies has been established in order to guide the implementation of the overall goals. These policies and business rules can be interpreted as architectural principles, allowing commanding units (the organizational echelons) to restrict the design freedom of their subordinates.

From the dynamic point of view, on-going flight operations, search and rescue missions, and defence readiness are controlled by feedback loop 1, representing the air force's daily business. The strategic initiatives leading to the formulation of the objectives model relate to feedback loop 2. They represent a planned change of the business, a new organizational configuration that is being designed (the equivalent EA term would be a to-be model) and that eventually replaces the existing configuration (the as-is model). In order to guide the change in organizational configurations, the Portuguese air force is developing a strategy map that represents the transition path considered most feasible to reach the given goals and objectives. Feedback loop 3 would be entered if the organization was faced with extreme environmental turbulences, such as a coup d'état.

The case study particularly points out the importance of organizational self-awareness, i.e. the idea that in order to perform any kind of control, system variables need to be observed. To this end, a wide array of indicators and dashboards has been developed to provide live information on the organizational status. These indicators cover different levels of description, i.e. the information contained in them describe the overall system on different strata, from flight operations of individual squadrons to overall organizational issues like budget availability for certain aircraft or technology components. Concrete steps towards increasing organizational self-awareness include the introduction of a formally agreed-upon terminology throughout the air force, moving from paper-based to electronic document storage and retrieval in order to increase dissemination and availability of information, uniform definition of roles (qualification, authority and responsibility) across the organization, and the continuing introduction of metrics. Metrics can be used to demonstrate to individual actors their contribution to the overall goals and objectives of the air force, as well as to observe more system variables and thus reduce the latency of the feedback loops.

5   Discussion

As discussed in section 3, the notion of feedback is central to both running the business (feedback loop 1) and changing the business (feedback loop 2). Only when system variables are observed can they be controlled. In the case of EAM, the architecture team may supply models to describe the system on different strata, but the contents of these models are provided by all organizational actors. This is because organizational actors must be provided with a degree of freedom to act within their domains, otherwise an enterprise would lose the ability to react to exceptions. EAM therefore depends on the existence and use of feedback channels that allow each organizational actor to act as observers, detecting model changes within their domain and propagating this data into models. Therefore, model updating mechanisms are central to EAM. If the observer or modeller part in any feedback loop fails, then the system cannot be purposefully controlled or transformed. A concrete example of an update mechanism is provided by Castela et al. [24].

In the case of EA principles, there is also interplay between coordination from higher echelons to lower echelons, and feedback in the other direction. Existing EA principles (i.e., principles based on existing models) can be used to run feedback loop 1. Known exceptions that might occur in this loop can be dealt with within existing principles. However, when feedback loops 2 or 3 are entered, information must be fed back in order to adapt principles to the new environmental situation. For example, a principle in a bank stating that all development activities are to be performed using a certain programming language works well as long as the environment remains stable and the enterprise remains in feedback loop 1. However, when transformation occurs, and a banking application has to be offered on mobile devices that cannot be supported by the existing programming language, these principles need to be adapted to the new model [25]. While all organizational actors are responsible for adhering to principles in their domain, they also need to challenge the validity of existing principles when the environment changes (i.e., when feedback loops 2 or 3 are triggered). Therefore, organizational actors on all echelons take part in governance.

The application of control theory adds semantics to the understanding of the enterprise that is not covered by the original three dimensions of the hierarchical systems theory. The commonalities between all notions of hierarchy identified in section 2 also hold for a hierarchy of feedback loops:

(1) Higher feedback loops are concerned with larger aspects of the overall system than lower feedback loops. Feedback loop 1 deals with running the business. Many instances of feedback loop 1 run in parallel, but each one is performed for smaller chunks of the system. Feedback loops 2 and 3 are triggered by exceptions that concern larger system aspects.

(2) Running frequency decreases from feedback loop 1 to feedback loop 3. While feedback loop 1 runs permanently, feedback loops 2 and 3 are triggered by unknown exceptions in the environment and therefore run less frequently. Feedback loop 3 is triggered by unpredictable, sudden exceptions and requires improvisation. This occurs less frequently than predictable exceptions leading to purposeful transformation in feedback loop 2. The decrease in running frequency as one moves up in levels is also emphasized by the fact that previously unknown exceptions are added to the pool of known exceptions after they have successfully been handled.

(3) Problem descriptions on higher feedback loops are harder to formalize and contain a greater degree of uncertainty than problem descriptions on lower feedback loops. This is also consistent with the concept of rising environmental turbulence triggering unexpected exceptions and therefore being addressed by higher feedback loops.

We therefore conclude that feedback loops extend the three original dimensions of the theory of hierarchical systems as a fourth dimension. This fourth dimension forms a vertical, hierarchical decomposition of downward control and upward feedback flows in the system, analogously to strata focusing on problem description and layers focusing on goal decomposition. Higher levels in the strata dimension imply higher degrees of abstraction. Higher levels in the layer dimension imply less formalized decision goals. Higher levels in the feedback loop dimension imply a rising degree of environmental turbulence that leads to the occurrence of new, unknown exceptions in the enterprise. As a consequence, we formulate the following proposition:

P1: The three feedback loops specified in section 3 can be regarded as levels of an additional hierarchy dimension in the sense of Mesarovic's theory, with feedback loop 1 occupying a lower level and feedback loop 3 a higher level.

The dimension of feedback loops is orthogonal to the other three dimensions, in that it can be applied to the overall system or to certain subsystems, as can the other hierarchical notions of strata and layers [6]. Organizational actors as decision units on a given echelon run through this hierarchy of feedback loops to address problems in their domain. In a similar fashion, they use a number of layers to break down their decision problems into sub-goals and a number of strata or describe it on various levels of abstraction. Feedback loops add dynamic aspects, describing an enterprise at runtime that handles environmental influences. The original dimensions of hierarchy: strata, layers, and echelons concern static system aspects. Compare feedback loops to the governance hierarchy: While governance describes a static allocation of decision rights, authority and responsibility, feedback loops show how different governance entities interact and self-activate during different degrees of environmental turbulence. Regarding both static and dynamic system aspects is consistent with a view on EAM as not merely a passive, documentation-oriented exercise, but as an ODE approach that actively support and guides organizational design.

6   Related Work

Fundamental to the application of both the theory of hierarchical, multi-level systems and control theory is the conceptualization of enterprises as dynamic systems. For modeling static aspects of enterprises, modeling approaches such as DEMO focus mostly on the stratum dimension of enterprises, and methods like UML lack a theoretical foundation. As we are specifically interested in providing a clear distinction between different types of hierarchy, we chose the theory of hierarchical, multi-level systems to describe the composition of an enterprise.

For modeling dynamic aspects of enterprises, there are several approaches: For example, the Deming cycle [26], a four-step iterative management method, consisting of the phases of plan-do-check-act. This cycle can also be seen as a sequence of feedback loops (observe-model-control).

The Dynamic capabilities framework [21, 27] emphasizes a reconfiguration of an enterprise's operational structure in order to achieve a better fit with the environment. Dynamic capabilities such as business process management, research and development, or EAM all emphasize planned, structured transformation (contrasted to spontaneous, improvisational transformation), which is why maintaining a dynamic capability in an enterprise is associated with additional effort [28].

Beer's viable systems model (VSM) [29] describes autonomous systems that are able to survive in a changing environment. It can be used to describe how enterprises are affected by and react to environmental (i.e., market) dynamics. The VSM consists of five levels, where the first three levels (primary activities, communication, and control) are responsible for running the business, and the last two levels (environmental scanning and policy decisions) are responsible for changing the business.

We chose control theory because it provides the ability to model the connections between running and changing the business – optimization versus transformation – via multiple levels of connected feedback loops, and because of its fit with the theory of hierarchical, multi-level systems.

7   Conclusion

In this paper, we have provided a theoretical grounding for enterprise transformation on multi-level systems theory and control theory. We have also conceptualized a multi-level structure of feedback loops that may be regarded as a fourth dimension of hierarchy in addition to strata, layers and echelons. The concept of a fourth dimension is a proposition that needs to be further researched; in particular, it needs to be contrasted against other possible dimensions of hierarchy. We have further illustrated the importance of feedback channels using a concrete example of a transformation-supporting management approach, namely EAM. The main limitation of this paper is that it is mostly conceptual. The conceptualizations, especially the multi-level feedback loops and their EAM implications need to be further analysed and transformed into useful innovative artefacts in future work. This could for example be performed by case studies in active enterprise transformation projects, or by reviewing evidence from past transformation. Nevertheless, this work provides a conceptual grounding of enterprise transformation on two well-established theories that can be used to derive requirements for concrete artefact construction.

Acknowledgement. We thank Antonia Albani for reviewing the manuscript and her valuable comments. This work has been supported by the Swiss National Science Foundation (SNSF).

References

1. Rouse, W.B.: A Theory of Enterprise Transformation. Systems Engineering 8(4), 279–295 (2005)
2. Dietz, J.L.G., Hoogervorst, J.A.P.: Enterprise ontology in enterprise engineering. In: Proceedings of the 2008 ACM Symposium on Applied Computing, Fortaleza, Ceara, Brazil (2008)
3. Harmsen, F., Proper, H.A.E., Kok, N.: Informed Governance of Enterprise Transformations. In: Proper, E., Harmsen, F., Dietz, J.L.G. (eds.) PRET 2009. LNBIP, vol. 28, pp. 155–180. Springer, Heidelberg (2009)
4. Ross, J.W., Weill, P., Robertson, D.C.: Enterprise Architecture as Strategy. Creating a Foundation for Business Execution. Harvard Business School Press, Boston (2006)
5. Tamm, T., Seddon, P.B., Shanks, G., Reynolds, P.: How Does Enterprise Architecture Add Value to Organisations? Communications of the Association for Information Systems 28, 141–168 (2011)
6. Mesarovic, M.D.: Multilevel systems and concepts in process control. Proceedings of the IEEE 58(1), 111–125 (1970)
7. ISO/IEC/IEEE: Systems and software engineering – Architecture description (ISO/IEC/IEEE 42010:2011) (2011)
8. Radeke, F.: Toward Understanding Enterprise Architecture Management's Role in Strategic Change: Antecedents, Processes, Outcomes. In: Proceedings of the 10th International Conference on Wirtschaftsinformatik, WI 2011, pp. 497–507 (2011)

9. Aier, S., Gleichauf, B., Winter, R.: Understanding Enterprise Architecture Management Design – An Empirical Analysis. In: Proceedings of the 10th International Conference on Wirtschaftsinformatik, WI 2011, pp. 645–654 (2011)
10. Strano, C., Rehmani, Q.: The Role of the Enterprise Architect. International Journal of Information Systems and e-Business Management 5(4), 379–396 (2007)
11. Winter, R.: Organisational Design and Engineering - Proposal of a Conceptual Framework and Comparison of Business Engineering with other Approaches. International Journal of Organizational Design and Engineering 1(1&2), 126–147 (2010)
12. The Open Group: TOGAF Version 9.1 (2011)
13. Dietz, J.L.G.: Architecture. Building strategy into design. Academic Service, The Hague (2008)
14. Österle, H., Brenner, W., Hilbers, K.: Total Information Systems Management - A European Approach. John Wiley & Sons, Chichester (1993)
15. Åström, K.J., Murray, R.M.: Feedback Systems: An Introduction for Scientists and Engineers. Princeton University Press (2008)
16. Wholey, D.R., Brittain, J.: Characterizing Environmental Variation. The Academy of Management Journal 32(4), 867–882 (1989)
17. Child, J.: Organizational Structure, Environment and Performance: The Role of Strategic Choice. Sociology 6(1), 1–22 (1972)
18. Aveiro, D., Silva, A.R., Tribolet, J.: Towards a G.O.D. Organization for Organizational Self-Awareness. In: Albani, A., Dietz, J.L.G. (eds.) CIAO! 2010. LNBIP, vol. 49, pp. 16–30. Springer, Heidelberg (2010)
19. Páscoa, C., Aveiro, D., Tribolet, J.: Organizational Configuration Actor Role Modeling Using DEMO. In: Proper, E., Gaaloul, K., Harmsen, F., Wrycza, S. (eds.) PRET 2012. LNBIP, vol. 120, pp. 18–47. Springer, Heidelberg (2012)
20. Eisenhardt, K.M., Martin, J.A.: Dynamic Capabilities: What are They? Strategic Management Journal 21(10/11), 1105–1121 (2000)
21. Pavlou, P.A., El Sawy, O.A.: The "Third Hand": IT-Enabled Competitive Advantage in Turbulence Through Improvisational Capabilities. Information Systems Research 21(3), 443–471 (2010)
22. Argyris, C., Schön, D.A.: Organizational learning: A theory of action perspective. Addison-Wesley, Reading (1978)
23. Páscoa, C., Tribolet, J.: Organizational and Design Engineering of the Operational and Support Components of an Organization: The Portuguese Air Force Case Study. In: Harmsen, F., Proper, E., Schalkwijk, F., Barjis, J., Overbeek, S. (eds.) PRET 2010. LNBIP, vol. 69, pp. 47–77. Springer, Heidelberg (2010)
24. Castela, N., Zacarias, M., Tribolet, J.: PROASIS: As-Is Business Process Model Maintenance. In: Harmsen, F., Grahlmann, K., Proper, E. (eds.) PRET 2011. LNBIP, vol. 89, pp. 53–82. Springer, Heidelberg (2011)
25. Buckl, S., Matthes, F., Roth, S., Schulz, C., Schweda, C.M.: A Conceptual Framework for Enterprise Architecture Design. In: Proper, E., Lankhorst, M.M., Schönherr, M., Barjis, J., Overbeek, S. (eds.) TEAR 2010. LNBIP, vol. 70, pp. 44–56. Springer, Heidelberg (2010)
26. Deming, W.E.: Out of the Crisis. MIT Press, Cambridge (1986)
27. Teece, D.J., Pisano, G., Shuen, A.: Dynamic Capabilities and Strategic Management. Strategic Management Journal 18(7), 509–533 (1997)
28. Zollo, M., Winter, S.G.: Deliberate Learning and the Evolution of Dynamic Capabilities. Organization Science 13(3), 339–351 (2002)
29. Beer, S.: The Viable System Model: Its Provenance, Development, Methodology and Pathology. The Journal of the Operational Research Society 35(1), 7–25 (1984)

Identifying Combinatorial Effects
in Requirements Engineering

Jan Verelst[1], Alberto Rodrigues Silva[2], Herwig Mannaert[1],
David Almeida Ferreira[2], and Philip Huysmans[1]

[1] Normalized Systems Institute
Department of Management Information Systems
University of Antwerp
Antwerp, Belgium
[2] Department of Computer Science and Engineering
IST & INESC-ID
Lisbon, Portugal
{jan.verelst,herwig.mannaert,philip.huysmans}@ua.ac.be,
{alberto.silva,david.ferreira}@inesc-id.pt

Abstract. There are several best practices and proposals that help to design and
develop software systems immune (to some extent) to combinatorial effects as
these systems evolve. Normalized Systems theory, considered at the software
architecture level, is one of such proposals. However, at the requirements engineering (RE)-level, little research has been done regarding this issue. This paper
discusses examples related with this problem considering two distinct RE abstract levels, namely at the business and system levels. The examples provided
follow the notations and techniques typical used to model the software system
at such levels, namely DEMO/EO, BPMN, and UML (Use Cases and Class diagrams). The analysis of these examples suggests that combinatorial effects
can be easily found at these different levels. This paper also proposes a research agenda to further investigate this matter in terms of the effects of combinatorial effects, and envisions the mechanisms and solutions for dealing with
them. It is suggested that an artifact-based, domain-specific approach is best
suited to achieve highly agile enterprises and RE-processes in the future.

Keywords: Requirement engineering (RE), requirements specifications, combinatorial effects (CE), normalized systems.

1   Introduction

A software requirements specification is a document that describes multiple technical
concerns of a software system [1,2]. A requirements specification is used throughout
different stages of the project life-cycle, namely to help sharing the system vision
among the main stakeholders, as well as to facilitate their communication, the overall
project management, and system development processes. A good requirements specification provides several benefits, namely [7,19-24]: establishes the basis for agreement between the customers and the suppliers on what the system is expected to do;

H.A. Proper, D. Aveiro, and K. Gaaloul (Eds.): EEWC 2013, LNBIP 146, pp. 88–102, 2013.
© Springer-Verlag Berlin Heidelberg 2013

reduces development efforts; provides a basis for estimating costs and schedules; provides a baseline for verification and validation; facilitates the system deployment; and serves as a basis for future maintenance activities.

Over the past two decades, it has become clear that organizations are increasingly facing more volatile environments. However, there are many indicators that organizations typically find difficult to cope with these changes in terms of their information systems. For example, the high percentage of challenged or even failed IT-projects clearly illustrates this problem [28]. For better describing this situation, some authors have even coined the term "software crisis". However, *change* does not only affect software, more specifically information systems. The respective requirements specifications are affected as well. Moreover, requirements specifications are the earliest documents in the systems development life cycle, and thus one of the first artifacts to be affected by change. Therefore, requirements specifications should be capable of dealing with change, namely taking preventive measures in terms of their structure and content to avoid such changes from causing a ripple effect at subsequent software development phases, such as software design and implementation.

Normalized Systems (NS) theory is especially concerned in studying the behavior of modular structures, such as software architectures, under change [4,5]. From a systems theoretic perspective, this theory has shown that evolvability and flexibility are largely determined by the presence of combinatorial effects (CE). Such CEs can be regarded as a kind of coupling and, more specifically, a ripple effect that is independent from aspects such as programming languages, systems development methodologies or frameworks used. Furthermore, CEs exhibit a highly harmful characteristic: these effects grow as the modular structure grows larger, which commonly occurs in practice over time. According to these empirical observations, the behavior of CE correlates with Lehman's law of increasing complexity, which states that, as maintenance is performed on a software system, its structure degrades and becomes more complex, thus making it inflexible [6]. This way, the existence of CEs explains why and how Lehman's law occurs. Furthermore, NS theory suggests that studying evolvability, as well controlling CE, is a highly complex endeavor, as CEs can occur at many levels in information systems and software architectures, from high-level effects at RE-level to very detailed effects at the implementation level. Usually, most CEs can be found at the lower level, where the large amounts of (cross-cutting) concerns make it difficult to avoid them.

In this paper, we report on our experiences focused on CEs at RE-level. Our research was motivated by several goals. Firstly, RE is a crucial discipline to be performed at the beginning of the systems development process. The existence of CEs at RE-level indicates that these modular artifacts exhibit limited evolvability, regardless of the software systems derived from them. In general, this limited evolvability is problematic because of the considerable effort involved in the RE process, as well as the impact of its main delivery in terms of the remaining phases of the software development process. Additionally, it also negatively influences the motivation and ability of requirements engineers to update their artifacts over time, which can lead to misalignments between the requirements specification and the software where these changes were effectively applied. Given that requirements specifications are often the basis for complementary technical documentation about the information system, as well as part of legal documents surrounding the corresponding project (including RFP or Project Contracts), several problems can result from CEs at RE-level.

Secondly, the concept of evolvability at RE-level is often overlooked. For example, in object-oriented (OO) literature, it is sometimes assumed that RE is substantially based on anthropomorphism, in the sense that making models about the problem space mainly consists in passively identifying real world objects. This approach has been argued by authors such as Simsion, who claim that data modeling should be considered more a design activity than an analysis activity [27]. The presence of CEs and coupling favors the latter perspective, and suggests that the RE-process entails many more issues than following a passive, analysis-like identification of objects in the real world. Indeed, NS theory suggests that studying evolvability is a complex, multi-level approach that, in turn, suggests different and more complex approaches to study the evolvability at RE-level are still needed.

The paper is organized as follows. Section 2 introduces the NS theory and the constructs and models commonly used in requirements specifications. Next, Section 3 provides examples of CEs using different notations and techniques, such as DEMO, BPMN, and UML diagrams.   In Section 4, we argue that these examples illustrate the need for a systematic research agenda on the identification and control of CEs at RE-level. Finally, Section 5 presents our conclusions.

2   Background

This section introduces the NS theory and the constructs and models usually used at the requirements specifications level.

2.1   Normalized Systems Theory

Normalized Systems (NS) theory studies how modular structures behave under change [4,5]. Initially, this theory was developed by studying change and evolvability at the software architecture level, by applying concepts such as stability and entropy to the study of the modular structure of the software architecture. Considering the application of systems theoretic stability to software architecture, *stability* implies that a bounded input function should result in bounded output values, even as T→∞. In software architecture, this means that a bounded set of changes should result in a bounded amount of changes or impacts to the system, even for T→∞. The concept of stability warrants that the amount of impacts caused by a change cannot be related to the size of the system and, therefore, needs to remain constant over time as the system grows. In other words, stability demands that the impact of a change is only dependent on the nature of the change itself. If the amount of impacts is related to the size of the system, a *combinatorial effect* (CE) occurs.

Research has shown that it is very difficult to prevent CEs when designing software architectures. More specifically, it has been proven that CEs are introduced each time one of four theorems is violated. The first theorem, *separation of concerns* , implies that every change driver or concern should be separated from other concerns. Applying this principle prescribes that each module can only contain one submodular task (which is defined as a change driver), but also that the implicit workflow should be separated from functional submodular tasks. The second theorem, *data version transparency* , implies that data should be communicated in a version transparent way between components. This requires that this data can be changed (e.g., additional data can be sent between components), without having an impact on the components and their interfaces. The third theorem, *action version transparency* , implies that a component can be upgraded without impacting the calling components. The fourth theorem, *separation of states*, implies that actions or steps in a workflow should be separated from each other in time by keeping state after every action or step. This suggests an asynchronous and stateful way of calling other components.

The proofs of the theorems show that unless every theorem is adhered to at all times during maintenance, the number of CEs will increase, making the software more complex and less maintainable. This can only be avoided when software is developed in a highly controlled way, ensuring that none of these principles are violated at any point in the development process during development or maintenance, which is quite difficult to achieve in practice. A modular structure that is free from CE, is called a Normalized System (NS). In order to achieve this, CEs should not be present at compile time, deployment time, and run time in modular structures. Furthermore, it has been shown that software architectures without CEs can be built by constructing them as a set of instantiations of highly structured and loosely coupled design patterns (called *elements*), which provide the core functionality of information systems and are proven to be free of CE.

This approach allows considering these software patterns as reusable building blocks, which can be aggregated using a mechanism called *expansion* to build information systems based on these building blocks without introducing CE. This contributes to realizing the vision of Doug McIlroy, who hoped for a future for software engineering in which software would be assembled instead of programmed. It is important to note that such assembly requires modules which are purposefully designed to prevent CE. Only when the absence of CEs in every pattern has been confirmed, it is possible to reuse these patterns without consulting their internal construction. Putting it in other words, only then can they be regarded as black boxes for usage in information systems. The theorems and patterns are described in terms of modular structures, which are independent of a given programming language or paradigm. As a result, these theorems and patterns have a wide applicability. More importantly, this shows that, in order to identify CE, and prescribe guidelines to prevent them, a modular structure in the domain under investigation needs to be made explicit, and the reuse of the modules in a black-box way should be confirmed.

2.2   Constructs and Models in RE

Requirements specifications define a somehow rigorous set of statements that help sharing a common vision between business stakeholders and the development team, and facilitates the communication, negotiation and managing efforts among all involved stakeholders. In general, requirements are specified in natural language due to their higher expressiveness and ease of use [7]. However, the usage of unconstrained natural languages often presents some drawbacks such as ambiguity, inconsistency and incompleteness. To mitigate some of these problems, specifications in natural language are typically complemented by some sort of controlled or semi-formal language – usually graphical languages such as UML [8], SysML [9], i* [10] or KAOS [11] –, which address different abstraction levels and concerns. Usually requirements engineers consider two distinct abstraction levels when organizing and specifying requirements: business level and system level. At the business level they define the enterprise and business context, and also the purpose and general goals of the system; while at the system level they have to further detail the concrete technical requirements of the system.

The constructs considered *at business level* are commonly the *terminology*, the business *goals* that the system should satisfy, and the *stakeholders* that are the sources of these goals and requirements, but also *business processes* and *business use cases*. There are in the community some languages that address the design of *goal-oriented models*, namely i* and KAOS. There are also other approaches to describe the system scope at this level, namely UML [8,12], BPMN [13], and RUP business modeling [14]. Additionally, depending on the size and complexity of the systems in consideration, enterprise engineering (EE) approaches can also be adopted at this level, for example using languages such as DEMO [15] or Archimate [35].

On the other hand, the main models considered in requirements specifications *at the system level* are context models, domain models, functional requirements models, and quality-attributes models. *Context models* use constructs such as *system, subsystems, components, nodes, external actors*, and respective relationships such as communication, interoperation, decomposition or deployment. Some of the visual languages that can be used to represent context models are SysML Block diagrams, UML Deployment diagrams, Data Flow Diagrams (DFD) at the context level [18], or even informal Block diagrams.

Domain models use constructs like *entities* or *classes*, and respective relationships such as associations and generalizations, and help to capture the key concepts or information resources underlying the system. The common graphical languages used to produce domain models are ER (Entity-Relationship) diagrams [18] or UML Class diagrams.

Functional requirements models use constructs such as *actors, functional requirements, use cases, scenarios* or *user stories*. There are different approaches to specify functional requirements. Most of these approaches recommend the use of textual specifications, written according to linguistic patterns properly enriched with predefined metadata and classifiers, such as priority and risk levels, authors, or creation dates. Other approaches recommend simple graphical representations such as UML Use Case diagrams or SysML Requirements diagrams. Yet, others recommend hybrid approaches by combining textual and graphical descriptions.

The concept of *non-functional requirements* (NFR) corresponds to high-level business constraints, technical constraints, and quality attributes [16,17]. Usually *business constraints* (e.g., a constraint related to the budget or the schedule of the project) are business level NFR but not included in requirements specifications because they used to be defined in other documents such as Project Charter and Project Plan documents. On the other hand, *technical requirements* (e.g., a constraint related the use of a specific development tool, the use of a particular database management system, or the adoption of a particular software development process) and *quality attributes* are considered system level NFR. *Quality-attribute models* use constructs like *qualities, metrics* and *utility values* to specify transversal properties of the system, such as maintainability, usability, performance, security, privacy or scalability. There are also some approaches to specify these requirements, namely the quality-attributes scenarios [17], or simple lists of quality-attributes [21]. Although quality-attributes are not difficult to be identified, they are hard to quantify in a verifiable manner. Since they can have a huge impact on the overall cost of the solution, they must be properly considered at the software architecture level [17].

3   Identifying Combinatorial Effects at RE Level

In this section we provide some illustrative examples of CEs that exist at requirements specifications based on the discussion of some notations and techniques commonly used, namely based on DEMO/EO, BPMN, and UML (in particular based on its Use Cases and Class diagrams). However, we start by discussing whether CEs can exist at the enterprise level (real world level), *irrespective* from these RE techniques.

Table 1. Analysis of Languages used in RE regarding Modularity and Combinatorial Effects

Abstract Levels	Approaches	Languages	Concepts	Relationships	Decomposition
			from Real World ...		
Business	Enterprise Ontology	DEMO	Service, Transaction, Act, Actor Role	Communication, Coordination, Production	Services compounded of transactions, transactions compounded of acts
	Business Processes	BPMN	Process, Resource	Control flow, Data flow	Process decomposition
System	OO System Analysis	UML Class Diagrams	Class	Association, Generalization	Class aggregation and composition
		UML Use Cases Diagrams	Use Case, Actor	Include, Extend	-
			... into Software Systems		

Table 1 summarizes the key aspects discussed below. As referred in section 2.2, requirements specifications can be defined at two distinct and complementary abstraction levels: business and system levels. The models produced at these abstraction levels can be somehow classified as those used in MDE (Model Driven Engineering) paradigm [36]. For example, considering the OMG MDA (Model Driven Architecture) approach, they can be classified, respectively, as Computational Independent Models (CIMs) and Platform Independent Model (PIMs). As it is expected, there are not Platform Specific Models (PSMs) defined at the RE level. We start by discussing the use of DEMO/EO and BPMN at the business level and then the UML (class and use cases diagrams) at system level. However, we do understand that because UML and (in somehow) DEMO are general-purpose modeling languages they could be used at both abstract levels.

3.1   from the Real World…

In information systems literature, it is commonly assumed (at least to a certain extent) that the information system should mirror the real world [26,27], which is also suggested by the concept of *anthropomorphism* that is frequently cited in object oriented literature. Together with communications theory-based approaches, such as DEMO, this would suggest that the real world is first and foremost an area of human behavior, which should therefore not predominantly be studied by theories based on computer science and/or automation. We agree with this point of view. Nevertheless, in modern society, human behavior increasingly takes place in highly structured, process-based contexts. Therefore, we argue that it is relevant to study these aspects of reality based on concepts such as *modularity*, while at the same time making an *abstraction* from purely human and communication aspects.

Therefore, an initial area for applying NS theory is the real world being mirrored. In other words, the first possibility is to investigate whether the real world itself consists of modular structures that are inherently unstable from a system theoretic perspective. To illustrate this, we take a simple example of a completely manual information system (not automated at all) at a university where student marks have to be rounded. Suppose the university has the policy of rounding exam marks "to the nearest integer". The university has the option to ask all professors to perform this rounding (option 1), but also to ask the administrative exam secretariat to perform this duty (option 2). Suppose now that the university policy changes: following option 1, the change impacts the number of professors that have to be notified to change their behavior, which is related to the size of the university, thus emphasizing a CE. In option 2, only one actor needs to be notified (the exam secretariat), implying that just 1 (or a few) physical person(s) have to be notified, which is largely or fully independent of the size of the organization. Therefore, option 2 has no (or only a negligible) CE. We stress that this example focuses on a system with no automated processing involved. Therefore, there is no combinatorial effect in the automation, but in 'the real world'.

A second example of a CE in the real world concerns the traditional *versus* virtual mail distribution. In certain organizations, most employees are entitled to write (physical) letters to external stakeholders. However, the logo's and letterheads of organizations are these days frequently changed, resulting in different paper and envelopes being used, and in this scenario, impacts every (secretary of) letter authors. This impact is dependent on the size of the organization, thus emphasizing another CE. Increasingly, organization are virtualizing their letters, by having authors send electronic versions of their letters to an internal or external mail center, who prints them, puts them in envelopes and dispatches them. In this second scenario, only one part of the organization is affected by the change of a company logo and letterhead, and therefore, no CE is present, or only an inconsequential one.

Both examples illustrate the existence of CE, without or prior to the use of RE techniques or notations, suggesting that they exist in the "Real World". Such CEs are (in a certain sense) outside the scope of the requirements engineer, as it is up to the business stakeholders to decide how to structure their organization and business processes.

3.2   DEMO/EO

An approach to RE is to start from enterprise models in order to give a high-level view of the business, and technical context of the system-of-interest. Among other alternatives, DEMO [15] have been used to support this goal, as well as a starting point for deriving use cases that describe the system functionality [29]. This is interesting for our approach, since DEMO models may be considered to be appropriate for analyzing CEs for the following reasons.

First, DEMO claims to create *constructional models*, instead of functional models. Constructional models represent the actual components of which a system consists. In contrast, functional models do not represent system components, but describe instead how a stakeholder uses the system. Possibly, this distinction explains why in Section 3.4   no CE could be identified: functional models do not consider the (modular) structure of a system, which was considered to be a prerequisite for identifying CEs in Section 2. In contrast, modular discussions based on DEMO models have already been described: for example Op't Land [30] argues that cohesion and coupling between actors in DEMO models can be used to decide whether or not to keep organizational actors together when splitting organizations.

Second, DEMO explicitly considers organizational building blocks, and prescribes rules for their aggregation. Acts are considered to be the basic building blocks (i.e., atoms), which are combined to create transactions (i.e., molecules). In order to deliver services to the environment, collections of transactions are invoked (i.e., fibers). The composition axiom structures how transactions can be interrelated. Transactions are either (1) initiated externally, (2) enclosed, or (3) self-initiated. Therefore, the aggregation of transactions needs to occur in certain ways. NS theory shows that CEs are often introduced when aggregating such constructs, and that prescriptive guidelines are required to show how building blocks can be aggregated without CE. Because DEMO models have clearly defined building blocks and aggregation guidelines, an analysis of the attention given to CEs on this level could be feasible. To the best of our knowledge, it has not been researched yet whether eliminating CEs has been taken into account in the DEMO guidelines.

Third, it is at least remarkable that certain concepts from NS theory are similar to EO [31]. For example, consider the *separation of states* theorem. It states that "the calling of an action entity by another action entity needs to exhibit state keeping in normalized systems" [4]. Therefore, it prescribes how action elements can interact. This impacts, for example, the workflow element, which aggregates action elements. A workflow can reach different states by performing state transitions. A state transition is realized by an action element. The successful completion of that action element results in a defined life cycle state. The workflow specification determines which state transitions can be made. Similarly, the state of a transaction in EO is determined by the successful performance of acts. The result of such an act results in the creation of a defined fact. Despite the different terminology, a clear resemblance between NS and EO emerges: state keeping is enforced in NS theory by defining states, and in EO by creating facts. These NS states are the result of executing actions, whereas the EO facts are the result of executing acts. The set of actions that can be performed is determined by state transitions in NS, and occurrence laws in EO. While we do not claim the adherence of DEMO models to the *separation of states* theorem, it is remarkable that such similar concepts are implicitly used, specially considering the different theoretical background of both approaches (i.e., language-action perspective and systems theoretic stability, respectively).

Notwithstanding these arguments, it should be noted that many real world aspects cannot be represented in DEMO models, since they are implementation-independent. Consider for example the "round to the nearest integer" (example described above). Using DEMO models, no difference between the two situations could be determined: at most, the rounding is an action rule for a certain execution act. Who applies this action rule is not modeled: the person fulfilling the executor actor role for this transaction (e.g., the examiner) can apply it, or it can be delegated (to the exam secretariat). Therefore, certain CEs of the implementation in the real world will not be visible in DEMO models.

3.3   BPMN

BPMN [13], like other notations (e.g., UML Activity diagrams), allows modeling business processes and, hence modeling the business context of the system in consideration. BPMN provides constructs such as process, task, role, resources, and so on, and also relationships such as control- and data-flows. In BPMN, processes and tasks can be considered and analyzed as modular structures. Research has already identified CEs in business processes, and provided guidelines to prevent them [32]. As such, changes to a certain process will only need to be applied in a single process model, instead of in every model where the functionality of that process is needed.

For example, consider a payment process. A payment process constitutes a different concern than the business process that handles, for instance, a purchase order. Based on the *separation of concerns* principle, this functionality should therefore be isolated in a dedicated process. If the payment functionality is modeled in every business process requiring a payment, each of these processes would have to be able to capture every possible change in the payment functionality. For example, when cash payments are no longer allowed, or validating an e-banking transfer with a first-time customer. However, if the payment concern is isolated in its dedicated business process, only the dedicated payment process needs to be changed. All the fault handling regarding transactions is included in this process. As a result, a reusable process can be modeled. Any business process requiring a payment, can request an execution of the payment process.

Based on the *separation of concerns* principle, a set of 25 guidelines has been proposed to eliminate CEs in business process models. Each guideline starts from the identification of a possible CE, and prescribes a solution to prevent that CE. As such, these guidelines are less general than the NS theorems. Rather, these principles apply the NS theorems on the business process model level. Although requirements are not expressed using process models, these guidelines illustrate how CEs can be identified and prevented at the business level of requirements specifications.

3.4   UML Use Cases

Use cases are highly popular, detailed and semi-formal descriptions of functional requirements. By far, UML Use Cases diagrams are the most popular graphical representations of the system from its functional point of view. These diagrams depict the actors (i.e., end-users and external systems) that interact with the system through a well-defined number of use cases. Use Cases are related among themselves through include or extend relationships. In the end, use cases are described textually, and are therefore typically situated close to the real world-level.

On one hand, use cases do have some modular characteristics, namely: (1) the name of the use case can be considered a primitive form of interface; (2) pre- and post-conditions can also be considered to delineate the functionality of the use case, and therefore be considered part of the interface (more specifically, another use case can treat this use case as a black box, providing the functionality described in the post conditions); and (3) the workflow of the use case can be considered the content of the module.

To a certain extent, this allows the identification of potential CE. For example, the principle of *separation of concerns* can be applied to (groups of) steps in the workflow. Typically, this principle is violated when several Use Cases describe the same functionality, or even terminology in a redundant way. If such a redundancy does not exist "in the real world", but does exist in the Use Case, it is a CE at the Use Case-level, caused by the text-based constructs used in Use Cases. Other constructs may help to prevent such CE: for example, *tagging* parts of a workflow or individual user interface requirements could be used, to provide a hypertext-like structure, which supports the identification of the impact of certain CE, and perhaps even the reduction of the number of impacts. Even though hypertext and tagging have limitations in terms of coupling in modular structures, they at least provide a better structure than plain text to judge the presence of CE.

However, textual descriptions have severe limitations in terms of CE. Use Cases are usually too underspecified to allow thorough identification of CE. For example, *action version transparency*, *data version transparency*, and *separation of state* theorems can be applied to the module interface (describing when Use Cases call each other). In Use Cases, however, this is difficult to judge because no interface parameters are detailed. Also, Use Cases give virtually no guidance as to which concerns should be separated, and therefore they are prone to scatter certain concerns over the entire document. In turn, this can lead to mixing functional with non-functional concerns, as well as mixing several non-functional concerns. For example, Use Case documents could contain non-functional user interface details in many or every Use Case (describing functional concerns). A change in the user interface requirements may then require a very large number of updates across the entire document.

This under-specification and lack of guidance is typically pointed out as one of the criticisms of EE (Enterprise Engineering) researchers, regarding Use Cases. Indeed, identifying CEs can be done in a more precise way in EE-approaches, such as DEMO.

3.5   UML Object-Oriented Domain Models

As mentioned in the previous section, domain models capture the key concepts of the system of interest. UML class models are the most popular notation for such domain models. Such models depict the classes and their relationships, such as associations and generalizations. Each class can both define data (attributes) and functions (methods) properties.

Concerning data, redundant definition of attributes are well-documented examples of CE, and the application of Codd's normalization rules [34] eliminates many of them. Concerning functions, the use of atomic data types in the interface of a method is a common violation of data version transparency. The CEs becomes clear when changing the definition of the attribute, which subsequently has to be applied to all redundant instances of the attribute. On the other hand, concerning the relationships between classes, the use of "sync pipelines" is an example of a violation of separation of state. This refers to a typical style in OO analysis, design and programming where method A calls method B, which calls method C, which calls method Z... while method A is still waiting for a return value from B. Similarly, X could call Y, who calls Z. In this case, the addition of one new error state in Z, would impact every calling method, i.e. both Y (and probably X), and C (and probably A and B).

These CEs are very similar to those in OO programming, which have been documented in [4].

4   Discussion

This discussion reflects our experience in the field both in practice and research. On one hand, we have more than 10 years' experience specifying as well assessing and auditing complex information systems based on different types of requirements specifications and related documentation. On the other hand, we have also researched in areas such as software architectures and system design (e.g., the Normalized Systems theory and its application [4,5]), requirements specification languages (e.g., the ProjectIT-RSL [23] or RSLingo approaches [24]), and the alignment between RE and MDE fields [25]). The scope and contribution of this paper reflects this blended experience.

The examples of CEs mentioned in the previous section are relatively straightforward, but they are sufficient to illustrate the omnipresence of instabilities in a domain that is sometimes considered to be about "identification of objects in the real world".

Indeed, these examples illustrate that both the real world (enterprise itself and respective information systems) and the "mechanisms and tools" we use to model them (e.g., classes or informational entities, use cases, processes and workflows, enterprise ontologies and communication acts) contain these instabilities. All of these CEs will exhibit Lehman-like symptoms. Initially, when the system is small, they would probably not be problematic, but over time their effects would grow and slowly but surely increase the rigidity of requirements models and specifications (which are sometimes used as the technical documentation of the information system, or a component in a legal contract concerning the system).

As summarized in Table 1, the examples above illustrate the existence of CEs at different abstraction levels, from the "real world" to enterprise-level and business processes descriptions (such as DEMO and BPMN), to object-oriented UML diagrams that have similar constructs as the implementation levels of information systems. This suggests that the RE- and systems development process consists of bridging a set of functional/constructive gaps, where every constructive level realizes the functional requirements of the functional level above using its own constructs.

At every level of this set of functional/constructive gaps, a certain amount of control of CEs should be striven for. On the one hand, it is clear that it is advantageous to eliminate the CE, and that is what we advocate at the software level: maximal elimination of CE. However, we explicitly mention that this is not necessarily the case at higher levels. For example, at the enterprise level, it seems quite probable that a certain level of CEs can be tolerated (for example, an organization may decide to keep using physical letter distribution over a virtual mail system), but in any case this decision should then be taken in a conscious way. At the level of the RE-techniques, it seems more certain that one should strive for full control of the extra CEs that are incurred. For example, the coupling in text-based requirements specifications' should be investigated for additional CE, as well as coupling CEs in BPMN models [32].

It should be remarked that the examples shown above are relatively straightforward, and that an experienced RE-practitioner is probably currently able to deal with most of these CEs in a heuristic way (based on experience). However, at the software level with its high number of concerns and correspondingly complex modular structures, heuristics have shown to be insufficient to control the large number of highly complex CEs that are responsible for the symptoms of Lehman's law. The enterprise and its supporting information systems are widely assumed to become even more complex in the future. This perspective may imply enterprises becoming larger in terms of amounts of different products, markets and human resources, but also relatively small enterprises can be faced with very high levels of complexity as they are part of multi-actor value networks that grow in size and complexity.

As this process of complexity increase takes place, heuristics applied by individual members of a RE-team will increasingly fall short in controlling coupling at the RE-level, and the need for a systematic approach to dealing with CEs is increasingly needed. Such a systematic approach should address minimally the following issues:

First, identification of CEs at each level, both in the constructs of the level and the models built using these constructs (also other NS-related concepts such as entropy should be considered at each level, but this is outside the scope of this paper).

Second, different mechanisms to achieve control of CE, such as the code genera-tion or expansion mechanisms that was used at the software level [5], but perhaps manual or semi-automated mechanisms are more appropriate at higher levels.

Third, appropriate levels of control of CE, and extent to which they need to be ap-plied in different levels of circumstances. For example, as mentioned above, possibly the higher levels should have higher tolerance-levels for CEs than more im-plementation-oriented levels. It is also possible that different levels are required depending on the sector the enterprise is situated in (for example, the types of changes that occur at high frequency in a sector).

The combination of these three issues, suggests that a single abstract, domain-independent approach in unlikely to achieve this ambitious goal of building the agile enterprise of the future. It is more likely that domain-dependent approaches are needed to focus fully on the subtle and complex issues surrounding coupling at all different levels in a certain sector. This is similar to classical engineering where reusable, domain-specific artifacts are constructed in sector like computer hardware design, car manufacturing, etc. Coupling is in these approaches also addressed by splitting the problem of car manufacturing in a series of sub-problems, i.e. design and manufacturing of the engine, the dashboard etc. Such a domain-dependent approach would mean that loosely coupled artifacts need to be developed in areas such as finance, accounting, transport, human resources, or in subareas such as invoicing, staffing, project management, mail distribution, payments, etc. All of these transver-sal subareas contain highly complex coupling issues which can be addressed by de-veloping artifacts such as invoice lines, address validators, credit checkers etc. When these artifacts are developed using a modular structure which exhibits control of coupling issues (such as a low number of CE), they can be aggregated into higher-order structures such as an invoice. This example may be surprisingly uncompli-cated, but at this point in time, there is no accurate description of the modular structure of an invoice available in the scientific literature, which is (widely) used in practice. On the contrary, invoices are currently still defined in practice in product-dependent and/or heuristic way, with no explicit study or science-based control of their modular structure.

Therefore, we believe that RE and EE would benefit from a piecemeal and induc-tive research agenda that is allowed by these domain-specific and problem decompo-sition approaches, in order to perhaps generalize to domain-independent techniques and methodologies in the future. However, we remark that this approach contrasts with the current mainstream in RE-literature, where there is a focus on large numbers relatively domain-independent modeling languages (constructs), techniques, metho-dologies and tools being proposed, with limited systematic study of the characteristics of the artifacts that are constructed.

5   Conclusion

In this paper we have documented our experiences in looking for CEs at different levels in the RE-process. The examples cover CEs at the RE level based on the adoption of notations and techniques such as UML classes and use cases, DEMO/EO, and BPMN models. The examples presented are relatively straightforward, but enough to show the omnipresence of such instabilities in the RE levels. As a result, we have described the need for a research agenda focusing on the systematic research into CEs and related issues at the RE domain in order to build enterprises and their information systems that are able to exhibit new levels of agility that will be required in the future.

In this way, we support the call by Dietz et al. for the area of Enterprise Engineering to be developed [33]. The amount and complexity of issues that need to be solved to achieve the next generation of truly agile enterprises both in the service and industrial sector, both in the for-profit and not-for-profit sector, is such that a scientific basis focusing on structural issues (including coupling) will be required.

References

[1]   Pohl, K.: Requirements Engineering: Fundamentals, Principles, and Techniques, 1st edn. Springer (2010)
[2]   Sommerville, I., Sawyer, P.: Requirements Engineering: A Good Practice Guide. Wiley (1997)
[3]   Tun, T.T., Trew, T., Jackson, M., Laney, R., Nuseibeh, B.: Specifying features of an evolving software system. Software: Practice and Experience 39(11), 973–1002 (2009), doi:10.1002/spe.923
[4]   Mannaert, H., Verelst, J.: Normalized Systems: Re-creating Information Technology Based on Laws for Software Evolvability. Koppa (2009)
[5]   Mannaert, H., Verelst, J., Ven, K.: Towards evolvable software architectures based on systems theoretic stability. Software Practice and Experience (2012)
[6]   Lehman, M.M.: Programs, life cycles, and laws of software evolution. Proceedings of the IEEE 68(9), 1060–1076 (1980)
[7]   Kovitz, B.: Practical Software Requirements: Manual of Content and Style. Manning (1998)
[8]   Booch, G., Rumbaugh, J., Jacobson, I.: The Unified Modeling Language User Guide. Addison-Wesley (2005)
[9]   OMG, Object Management Group, Systems Modeling Language, http://www.omgsysml.org
[10] Yu, E.: Modelling Strategic Relationships for Process Reengineering, PhD thesis, University of Toronto, Canada (1995)
[11] Lamsweerde, A.: Requirements Engineering: From System Goals to UML Models to Software Specifications. Wiley (2009)
[12] Castela, N., Tribolet, J., Silva, A.R., Guerra, A.: Business Process Modeling with UML. In: Proceedings of the International Conference on Enterprise Information Systems. ICEIS Press (2001)
[13] OMG: Business process model and notation (bpmn), version 2.0. Tech. rep. OMG (2011)
[14] IBM Rational Method Composer and RUP on IBM Rational developerWorks, http://www.ibm.com/developerworks/rational/~products/rup/
[15] Dietz, J.L.G.: Enterprise Ontology: Theory and Methodology. Springer (2006)

[16] Chung, L., do Prado Leite, J.C.S.: On Non-Functional Requirements in Software Engineering. In: Borgida, A.T., Chaudhri, V.K., Giorgini, P., Yu, E.S. (eds.) Conceptual Modeling: Foundations and Applications. LNCS, vol. 5600, pp. 363–379. Springer, Heidelberg (2009)

[17] Bass, L., Clements, P., Kazman, R.: Software Architecture in Practice, 2nd edn. Addison Wesley (2003)

[18] Weaver, P., Lambrou, N., Walkley, M.: Practical SSADM Version 4+, 2nd edn. Prentice Hall (1998)

[19] IEEE, IEEE Std 830-1998 (Revision of IEEE Std 830-1993). IEEE Recommended Practice for Software Requirements Specifications (1998)

[20] Withall, S.: Software Requirements Patterns. Microsoft Press (2007)

[21] Robertson, S., Robertson, J.: Mastering the Requirements Process, 2nd edn. Addison-Wesley (2006)

[22] Cockburn, A.: Writing Effective Use Cases. Addison-Wesley (2001)

[23] Videira, C., Ferreira, D., Silva, A.R.: A linguistic patterns approach for requirements specification. In: Proc. 32nd Euromicro Conference on Software Engineering and Advanced Applications. IEEE Computer Society (2006)

[24] Ferreira, D., Silva, A.R.: RSLingo: An Information Extraction Approach toward Formal Requirements Specifications. In: Proc. of the 2nd Int. Workshop on Model-Driven Requirements Engineering (MoDRE 2012). IEEE Computer Society (2012)

[25] Silva, A.R., Saraiva, J., Ferreira, D., Silva, R., Videira, C.: Integration of RE and MDE Paradigms: The ProjectIT Approach and Tools. IET Software Journal 1(6) (2007)

[26] Borgida, A.: Features of languages for the development of information systems at the conceptual leve. IEEE Software, 63–72 (January 1985)

[27] Simsion, G., Witt, G.: Data Modeling Essentials, 3rd edn. Morgan Kaufmann (2004)

[28] Standish Group, The Standish Group Report: Chaos (1995)

[29] Shishkov, B., Dietz, J.L.G.: Deriving Use Cases From Business Processes, the Advantages of Demo. In: Proceedings of ICEIS 2003, pp. 138–146 (2003)

[30] Op 't Land, M.: Applying Architecture and Ontology to the Splitting and Allying of Enterprises, PhD Thesis, Technical University Delft (NL) (2008)

[31] Huysmans, P.: On the Feasibility of Normalized Enterprises: Applying Normalized Systems Theory to the High-Level Design of Enterprises, PhD Thesis, University of Antwerp (2011)

[32] Van Nuffel, D.: Towards Designing Modular and Evolvable Business Processes. PhD Thesis, University of Antwerp (2011)

[33] Dietz, J.L.G.: Enterprise Engineering Manifesto (2010),
http://www.ciaonetwork.org/publications/EEManifesto.pdf

[34] Codd, E.F.: A relational model of data for large shared data banks. Communications of the ACM 13(6), 377–387 (1970)

[35] Lankhorst, M., et al.: Enterprise Architecture at Work - Modelling. Communication and Analysis. Springer (2005)

[36] Stahl, T., Volter, M.: Model-Driven Software Development. Wiley (2005)

Understanding Entropy Generation during the Execution of Business Process Instantiations: An Illustration from Cost Accounting

Peter De Bruyn, Philip Huysmans, Herwig Mannaert, and Jan Verelst

Normalized Systems Institute (NSI)
Department of Management Information Systems
University of Antwerp
Antwerp, Belgium
{peter.debruyn,philip.huysmans,herwig.mannaert,jan.verelst}@ua.ac.be

Abstract. The instantiation and execution of business processes typically generates an enormous set of data, including financial- and accounting-related information, based on different aggregation levels. As a result, it can be very complex to draw conclusions from this data, such as which steps in a business process are causing delays or, in an accounting context, which tasks are causing high costs. In this paper, we relate this complexity generated through business process execution to the concept of entropy, as defined in thermodynamics. More specifically, we show how information aggregation seems to be at the core of this phenomenon. We discuss six types of information aggregation dimensions which tend to increase entropy (and hence, complexity) in a cost accounting context. As entropy is generally controlled by adding structure to the considered system, we propose a set of preliminary guidelines to control this entropy based on insights from the Normalized Systems (NS) theory rationale.

Keywords: Entropy, Business process execution, Information aggregation, Cost accounting, Normalized Systems.

1 Introduction

In order to make appropriate business decisions, managers require accurate information of the organization. For example, cost accounting approaches accumulate cost data to deliver precise information on the costs to design, produce and deliver certain products or services. However, various authors argue that even advanced cost accounting approaches (e.g., Activity-Based Costing) have issues to adequately report on complex and changing product portfolios [1], especially when such products are produced by complex processes [2]. As a result, it can be very complex to draw practical conclusions from this data, such as which steps in a business process are causing delays or, in an accounting context, which tasks are causing high costs. Also, a sound theoretical basis seems required to develop a suitable approach. In engineering sciences, complexity is

H.A. Proper, D. Aveiro, and K. Gaaloul (Eds.): EEWC 2013, LNBIP 146, pp. 103–117, 2013.

studied using the entropy concept. In this paper, we use entropy as defined in statistical thermodynamics to study the phenomenon of entropy generation during the execution of business processes. We will start by elaborating on some basic concepts related to the entropy concept in Section 2. Afterwards, we will show in Section 3 that entropy generation during the execution of business processes is due to the (uncontrolled or unconscious) aggregation of (cost) information according to several possible aggregation dimensions. The relevance of this analysis for the design of cost accounting systems in practice, and the information systems supporting these accounting systems will be discussed in Section 4.

It should be noticed that we are not the first to employ the concept of entropy to study the complexity of financial and accounting related data. For instance, Lev [3] proposed to analyze the design and information content of financial statements from an entropy viewpoint. Also here, the author concluded that improved decision making seems to be enabled by access to detailed, rather than aggregated, information. Later on, other authors have elaborated on this approach (see e.g., [4,5]). However, our approach seems to differ in several aspects. First, whereas Lev and the follow-up studies started their analysis from entropy as defined in information theory, we take the statistical thermodynamics perspective as our starting point. Second, whereas the mentioned studies mainly focused on financial reporting, our research is situated in a business process (i.e., managerial or operational) context (e.g., to perform business process optimizations). Third, our approach stresses more explicitly the importance of analyzing the run-time behavior of organizational constructs to introduce the concept of entropy.

2   Theoretical Framework: Entropy

As we use the theoretical concept of entropy in this paper to analyze the complexity generated through business process execution, we will introduce the necessary entropy concepts and definitions in this section.

Entropy as expressed in the second law of thermodynamics is considered to be a fundamental principle in traditional engineering sciences. While many versions exist, all approaches have basically the intent of expressing the (increasing) amount of complexity, uncertainty (i.e., lack of information) and tendency of particles to interact (i.e., couple) within a system. In this paper, we will use the perspective towards entropy as employed in statistical thermodynamics. Here, entropy was defined by Boltzmann as proportional to the number of possible *microstates* (i.e., the whole of microscopic properties of the particles of a system) consistent with a single *macrostate* (i.e., the whole of externally observable and measurable properties of a system), i.e., the *multiplicity* [6].

This definition of entropy can be further clarified by the following example. Consider a set of 100 coins, each of which is either heads up or tails up. Assuming that the observer is only able to consider the general outcome in terms of the total number of heads and tails, this information specifies the macrostate. The set of microstates is then specified by the possible configurations of the facings of each individual coin resulting in the observed macrostate. For the macrostate of

100 heads or 100 tails, there is exactly one possible configuration (i.e., all coins are heads or all coins are tails respectively), so our knowledge about the system is complete (i.e., multiplicity equals 1). At the opposite extreme, the macrostate which gives us the least knowledge about the system consists of 50 heads and 50 tails in any order, for which there are 10^{92} possible microstates (i.e., multiplicity equals 10^{92}). It is clear that the entropy is extremely large in the latter case because we have no knowledge of the internals of the system.

A common way of dealing with entropy, is to increase the *structure* or the knowledge of the internals of the system. Consider again our coin example. The entropy in this example can be reduced when we add structure to the studied system. Suppose we would create —instead of 1 group with 100 coins— 10 groups of 10 coins, each with 5 heads and 5 tails. In this situation, multiplicity would only amount to 2520 [7]. Consequently, the entropy for this system would be much lower. Structure can be used to control entropy, in the sense that by allowing less interaction between the constituting components before the information is observed, a lower number of valid combinations is possible. This leads to less uncertainty concerning the actual microstate configuration.

The mechanisms related to entropy reasoning have found their reflection in many domains, including business and management topics. Even for the business process management domain, some contributions can be found (see e.g., [8]). However, an approach based on entropy as defined in statistical thermodynamics for studying the complexity arising from the (cost) data generated by executing business process instantiations, is, to the best of our knowledge, non-existing.

3   Entropy Generation and Aggregation Dimensions in a Business Process Context

In this section, we will demonstrate how entropy generation in a business process context can essentially be explained by considering the aggregation of information during the execution of business process instantiations (i.e., considering the run-time environment) according to several possible aggregation dimensions. In order to do so, we will first discuss our conceptualization of a business process instantiation space and the corresponding definition of microstates and macrostates in such context. We then discuss a set of six possible (cost) information aggregation dimensions during the execution of business processes, how we can see these aggregation dimensions as different degrees of entropy, and explore how we can avoid this entropy generation.

3.1   The Run-Time Instantiation Space of a Business Process

Many (diagnostic) management decisions are typically related to the run-time perspective of business processes, as properties like a realized throughput time or the costs of specific instantiated products or services only exist in such run-time perspective (and not in a merely design-time perspective). Therefore, we need to define the necessary run-time instantiation space of business processes.

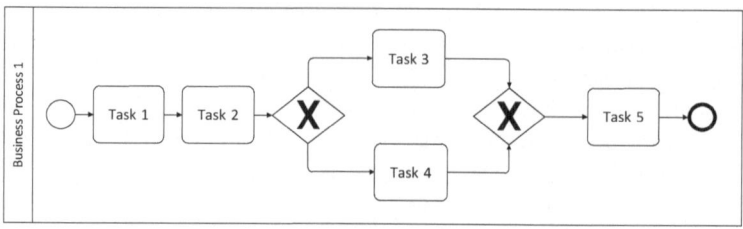

Fig. 1. A general business process BP_1 consisting out of five tasks

In order to do so, let us regard a business process as a flow which constitutes a set of consecutive tasks (including selections, iterations, etc.) on an information object (e.g., an invoice, the production of a car, etc.) with the intent to attain a certain business goal. Consider for instance the exemplary business process as depicted in Figure 1 (labeled "Business Process 1" or "BP_1" further on). The business process consists out of five tasks, of which tasks 3 and 4 are part of an exclusive gateway, meaning that only one of them will be executed during each instance:

$$BP_1 = \{t_1, t_2, t_3, t_4, t_5\}$$

Such business process might be imaginable for both industrial or more administrative purposes. For instance, in an industry context, the process of delivering a certain product might consist of 5 tasks performing respectively: assuring the client's creditworthiness (i.e., task 1); analyzing whether the requested good is still in stock (i.e., task 2); in case the product is in stock: retrieval of the product out of stock (i.e., task 3); in case the product is not in stock: product assembly (i.e., task 4); product shipping (i.e., task 5).

Each time a product is asked to be delivered , a new instance of the business process is initiated and an instantiated information object (i.e., a specific product delivery instance) passes through each step in the process. For the moment making abstraction of the groupings depicted by the dotted lines, an example of such business process instantiation space is provided in Figure 2.

As such, each *business process instantiation* $BP_{i,j}$ can be identified using the index i to refer to the business process type (here only "1" as we consider only one business process type) and index j to refer to the business process instance of a particular business process type (here instances "1", "2", and "3" of business process type "1"). At its turn, each business process instance contains a set of instantiations of its constituting tasks $t_{k,m}$. Again, each *task instantiation* can then be identified using indexes k for the task type (here tasks "1" to "5" contained in business process type BP_1) and m for to the task instance. Hence, the *business process instantiation space* in our example becomes:

$$\begin{cases} BP_{1,1} = \{t_{1,1}, t_{2,1}, t_{3,1}, t_{5,1}\} \\ BP_{1,2} = \{t_{1,2}, t_{2,2}, t_{4,1}, t_{5,2}\} \\ BP_{1,3} = \{t_{1,3}, t_{2,3}, t_{3,2}, t_{5,3}\} \end{cases}$$

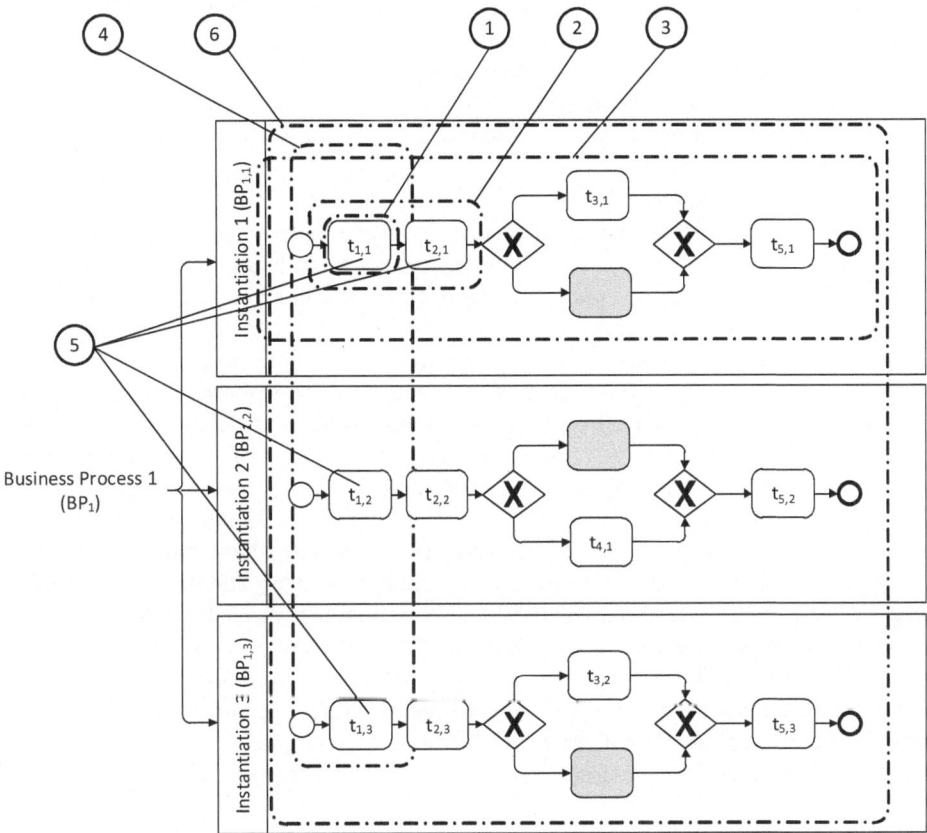

Fig. 2. A set of business process instantiations of "Business Process 1" and 6 possible information aggregation dimensions of the costs involved in each task instance

3.2   Interpreting Microstates and Macrostates

We discussed in Section 2 how entropy generation can essentially be traced to the degree of interaction and coupling between the particles or modules making up a system, during its lifetime. As such, for each modular system, entropy could be studied. In previous work, we therefore already briefly argued that entropy reasoning can be applied to organizational systems as well [9,10]. Accordingly, macrostates and microstates should and can be defined in such framework.

Aiming to employ the thermodynamics analogy, the individual task instantiations can be seen as the "particles" in the business process system. Whereas in typical thermodynamics, properties as speed and position of the particles are studied, its counterpart in the context of tasks would be typical properties like the throughput time of an individual task instantiation in a process, its correct or erroneous outcome, the costs and resource consumption of an individual task, etc. Hence, the *microstate* in the defined instantiation space is given by the union of the values of properties (e.g., costs) for each individual part (i.e., task instantiation): $\{C(t_{k,m})\}_{k,m}$. The *macrostate* of this space is the (aggregated) information available for the observer, generally entailing unrecoverable loss of information. Consequently, in a business process context, a macrostate could be considered as referring to typical observable information of business processes such as the total throughput or cycle time, quality or output measures, total costs, resource consumption, etc.

As in traditional thermodynamics, entropy is also here conceived as being proportional to the number of microstates consistent with a single macrostate (i.e., multiplicity Ω). In case we can now easily solve typical management questions of the type *"which task(s) was (were) responsible for the extremely large cost or extremely long throughput time of a particular business process (instance)?"*, entropy can be considered to be low. In such case, the observable macrostate (i.e., the extremely large cost or extremely long throughput time) can be related to only one or a few microstates (i.e., the responsible task or tasks). In contrast, when the answer to such questions is unclear (i.e., many tasks can possibly be responsible for the observed macrostate), entropy is high. In such case, one could argue that different pieces of information (i.e., the particles) "interact" or are being "coupled" in front of the observer, trying to identify the origin of a (problematic) observation. As a consequence, also in business process context, entropy can be seen as a measure for the lack of information one has on a system and hence, uncertainty. Therefore —making abstraction of the possible costs entailed in controlling occurring entropy— situations in which entropy is as low as possible, seem desirable. Based on these formalisms, we can now perform several analyses regarding information aggregation and entropy generation.

3.3   Possible (Cost) Information Aggregation Dimensions during Business Process Instantiation Execution

Typically, each of the steps (i.e., the instantiated tasks) in an instantiated business process can be associated with some costs. Such costs may originate from raw material consumption, electricity consumption, equipment usage, personnel load, etc. The specifics of this cost structure might obviously vary significantly from the product and process under consideration. For instance, in a more administrative process, the major emphasis may be on personnel load. Alternatively, typical industrial processes will probably have a focus on raw material, electricity and equipment usage.

Regardless of the specific cost structure, each process owner might be interested in how much it costs to produce a set of products or deliver a set of services for (for example) pricing and accounting purposes. In terms of efficiency optimization, it might even be necessary for the business process owner to know a more detailed breakdown of these costs and split them up according to each separate task for which the cost information is relevant (i.e., an *"information unit"*) and according to individual product or service instances. For instance, in case the process owner notices that a product (instance) has an extremely high cost, his first concern would probably be to locate (i.e., diagnose) which task or tasks were responsible for this high cost. Also in general process optimization efforts such as Business Process Reengineering (BPR) or Total Quality Management (TQM), such information might be highly valuable to purposefully direct one's attention for amelioration.

Nevertheless, both consciously and unconsciously, information regarding the execution of (tasks of) business process instantiations is frequently aggregated at a higher level than an individual instance of a task. This might be the case for several (sometimes appropriate) reasons, including measuring difficulties or the inherent costs or perceived overhead for registering the costs at this fine-grained level. Also, these aggregations can frequently be considered to be relevant in a business situation. We highlight six of such possible information aggregation dimensions:

Aggregation Dimension 1. Information is gathered and recorded at its most fine-grained level: for each individual instantiation of each business process type, the costs per task instantiation on an individual information object are recorded. For instance: the actor responsible for checking the completeness of a damage cost reimbursement request at an insurance company (i.e., one task or "information unit"), records his time spending for checking and completing each individual request in a spreadsheet. Indeed, the time (and hence: costs) an actor invests in checking such completeness might vary from very little time (e.g., a request which contains all relevant documents in the proper way) to a lot of time (e.g., a request in which many information is initially lacking) and might be dependent on both the instance level (i.e., the specific damage cost reimbursement request) and the business process type level (i.e., more complex requests requiring typically more time to check completeness). In fact, at this aggregation dimension, no aggregation or interaction with any other "information units" occurs.

Aggregation Dimension 2. Information regarding two or more "information units" k is aggregated within the scope of one single business process instance j. Such situation might be the case when only the information (on costs) for certain (major) phases in a production process is recorded. While an organization might be convinced that recording the information at a more fine-grained level seems unnecessary or not cost-effective, such aggregation might also be of interest for (external) stakeholders of the company. Indeed, in case of very complex business processes, one can imagine that clients or certain actors at a higher management level might be primarily interested in the mere "milestones" (e.g., "order received", "order produced", "order shipped") of a business process for monitoring purposes, instead of the possibly hundreds of more fine-grained states the product might be in during its lifecycle.

Aggregation Dimension 3. In fact, this aggregation dimension is a more general case of aggregation dimension 2. Here, cost information is aggregated over all tasks k per business process instantiation j (e.g., for each product assembled, a total assembly cost is available). Such aggregations are especially useful for, for instance, customized price settings in which a company might adopt the strategy to charge a client a price for a product or service based on the cost-plus pricing principle (i.e., $price = X \%$ $product$ $instance$ $cost$, where $X \geq 100$). Also, especially in case a company produces high-value or custom made products, this aggregation might be particularly valuable to comply with reporting and bookkeeping standards or regulations.

Aggregation Dimension 4. This dimension considers the aggregation of costs among all instances m of a particular task k within BP_i. Such situations are conceivable in case (when elaborating on the damage cost reimbursement request business process) a specific operator is solely put in charge of checking the completeness of damage cost reimbursement requests (i.e., task 1). In such situation, the cost for employing this person is to be divided over all task instances m of t_1 of the considered business process type BP_i. Also, while the analysis of an instantiation space for multiple product or business process types is out of scope for this paper, it is clear that this aggregation dimension can further be generalized over multiple business process types in case multiple business process types incorporate the same task t_k.

Aggregation Dimension 5. In this aggregation dimension, the (cost) information is aggregated according to the time elapsed. This means that costs are aggregated as time goes by until the observer stops the "counter" at a certain point in time t for further inspection. These aggregations primarily seem to occur in industrial settings. For instance, at a manufacturing plant, it seems reasonable to have cost information on, for example, electricity consumption in this way. In doing so, a counter recording the electricity consumed throughout time may offer an observer insight for the costs involved at each point in time t which is desired. However, no explicit breakdown according to tasks, instances, etc. is made.

Aggregation Dimension 6. Information regarding all (task) instances of the considered business process type becomes aggregated. In this aggregation dimension, no distinction between separate tasks and business process instantiations is made and solely the overall outgoing cash-flows and costs related to the business process type are considered. While possibly present at other aggregation dimensions as well, one can find typically at this aggregation level many management-oriented KPI's (Key Performance Indicators) and accounting ratio's related to the total revenue generation (per product type), total costs (per product type), profitability (per product type), number of items (per product type) sold, etc. Further, while the analysis of the instantiation space for multiple product or business process types is out of scope for this paper, it is again clear that aggregations taking into account several product types can be realistic as well (e.g., total revenue, costs and profitability over all product and business process types). Moreover, this aggregation dimension can be considered as a special case of aggregation dimension 5, when point in time t is chosen in such way that BP_i has completed the execution of all its instantiations j.

These aggregation dimensions are visually represented in Figure 2 by the groupings indicated by the dotted lines in which each aggregation dimension is attached with a number equal to the enumeration provided above.

Table 1. Illustration of the interaction of the (cost) information of $t_{1,1}$ with other task instantiation information, according to the six proposed information aggregation dimensions. For each aggregation dimension column, the x's show with which information, the (cost) information of $t_{1,1}$ is aggregated.

business process instantiation	task instantiation	cost (€)	aggregation dimension					
			(1)	(2)	(3)	(4)	(5)	(6)
$BP_{1,1}$	$t_{1,1}$	17.2	x	x	x	x	x	x
	$t_{2,1}$	5.6		x	x		x	x
	$t_{3,1}$	5.2			x			x
	$t_{5,1}$	4.6			x			x
$BP_{1,2}$	$t_{1,2}$	5.1				x	x	x
	$t_{2,2}$	4.3						x
	$t_{4,1}$	4.8						x
	$t_{5,2}$	5.6						x
$BP_{1,3}$	$t_{1,4}$	4.8				x	x	x
	$t_{2,3}$	6.0						x
	$t_{3,2}$	4.4						x
	$t_{5,3}$	4.8						x
Aggregated cost	(AC)		17.2	22.8	32.6	27.1	32.7	72.2
Expected cost	(EC)		5	10	20	15	20	60
Relative deviation	$(RD) = \frac{AC-EC}{EC}$		2.44	1.28	0.63	0.81	0.64	0.20

Further, Table 1 describes the instantiation space as depicted in Figure 2 in an equivalent way: focusing on task instantiation $t_{1,1}$, it lists for each aggregation dimension the different tasks (and hence information units) with which the cost information would interact (i.e., would be aggregated) before it is externally observed. The table also provides an exemplary cost overview per task instantiation (i.e., microscopic), as well as the information available for the process manager (i.e., the macroscopic "aggregated cost") in case each of the presented aggregation dimensions is considered. For instance, while in aggregation dimension 1 the cost information of task instantiation $t_{1,1}$ (cost = 17.2) is recorded individually (hence, aggregated cost = 17.2), this cost information is aggregated with task instantiations $t_{2,1}$ (cost = 5.6), $t_{3,1}$ (cost = 5.2) and $t_{5,1}$ (cost = 4.6) in aggregation dimension 3 (hence, aggregated cost = 32.6). We made the assumption that the expected value of the cost for each task instantiation is the same and equal to 5 (i.e., $EC(t_{1,1}) = EC(t_{1,2}) = \ldots = EC(t_{5,3}) = 5$), as this makes our further analysis regarding problem identifications (based on the respective dimension) more straightforward for the considered example.

3.4 Understanding Business Process Entropy Generation by Information Aggregation

As we stated earlier that entropy generation can essentially be traced to the degree of interaction and coupling between the particles making up a system, the

Table 2. Illustration of multiplicities for each of the considered aggregation dimensions

	multiplicity (Ω)	
	Example	General
Aggregation dimension 1	1	1
Aggregation dimension 2	2	# combined information units k
Aggregation dimension 3	4	# tasks k in BP_i
Aggregation dimension 4	3	# task instantiations m of t_k
Aggregation dimension 5	4	# tasks $t_{k,m}$ executed at point in time t
Aggregation dimension 6	12	$\left\{ \begin{array}{l} \text{depending on \# instantiations } j \text{ of } BP_i, \\ \text{\# tasks } k \text{ and instantiations } m \text{ in } BP_i, \\ (\text{\# business processes } i \text{ in the repository}) \end{array} \right.$

different aggregation dimensions discussed in Section 3.3 each have a different degree of entropy which can be calculated. In Table 2, the multiplicities (and hence entropy) of the different aggregation dimensions are listed for both our exemplary business process and its instantiation space (i.e., column "example"), as well as the more general case for any considered business process type and its instantiation space (i.e., column "general") . The table starts from the assumption that one observes a (problematic) macrostate (e.g., the costs for executing our considered business process are too high) and one wants to detect which individual task (or possibly which set of tasks) was responsible for this problematic situation. In case the macrostate is uniquely traceable to one individual task (or set of tasks), multiplicity amounts to 1 and, hence, entropy is minimal. When the macrostate is consistent with multiple microstates, entropy increases. For instance, in aggregation dimension 2, when a problematic macrostate arises, the observer is only able to trace the result back to the aggregation of task instantiations $t_{1,1}$ and $t_{1,2}$. As such, the multiplicity is at least 2 as one is unable to detect whether it is $t_{1,1}$ and/or $t_{1,2}$ which is responsible for the increased costs, and entropy is higher than in the previous situation.

Generally speaking, the six aggregation dimensions were ordered in ascending order of entropy. For instance, it is clear that aggregation dimension 1 provides the most fine-grained cost information possible, its information units interact the less with other cost information, and allows for unambiguous traceability from a macrostate (e.g., total cost for all $BP_{i,j}$) to one microstate (i.e., a properly available cost for each individual task and its instantiations). Conversely, aggregation dimension 6 is clearly the most coarse-grained aggregation imaginable as it purely reflects the a macrostate by itself. Nevertheless, the strict order between aggregation dimensions 3 till 5 may be variable, according to the specific values of parameters k, j, m and t.

The fact that the aggregation dimensions with a higher number, tend to have a higher degree of entropy, has its implications for business process optimization efforts as well. Considering for instance again Table 1, where the rows "Aggregated cost", "Expected cost" and "Relative deviation" constitute the observable (i.e., macroscopic) information available for the process manager. We can see from the detailed (i.e., microscopic) cost information that task instantiation $t_{1,1}$ can be deemed problematic as it exceeds its expected cost by threefold. All other task instantiations remain within an interval of maximum 20% deviation. Analyzing the situation of a process owner who is aiming to diagnose and solve any excessive costs occurring in business process instantiations, we can notice two effects.

First, as entropy increases, the relative deviation becomes smaller and hence, the problem becomes *less observable*. Indeed, for aggregation dimension 1, the relative deviation amounts to 2.44 and therefore clearly highlights that an irregular task execution has taken place. For aggregation dimension 6, the relative deviation only amounts to 0.20 as the extreme value of $t_{1,1}$ is compensated by the "normal" costs of the other tasks, and might therefore not necessarily alert the process owner that something has gone wrong.

Second, the *traceability* to the responsible task instantiation becomes more difficult as entropy increases. Supposing that in each of the aggregation dimensions, the process owner is aware that some irregularity has taken place, the correct diagnosis becomes more difficult as entropy increases. Indeed, for aggregation dimension 1, the values of the relative deviation unambiguously point to task $t_{1,1}$. In case of aggregation dimension 2, the attention of the observer is caught by the relative deviation concerning tasks $t_{1,1}$ and $t_{1,2}$. While this observation gives a clue to where the problem is situated, the observer should still investigate to which of the two considered task instantiations, the error can be attributed. However, in case of aggregation dimension 6, the observer is left with no real indication as to where precisely in the full instantiation space the problem can be situated. Therefore, he should scrutinize the operation of all task instantiations of all considered business process instantiations to find out.

3.5 Controlling Business Process Entropy by Increasing the Structure of the System

Generally speaking, entropy control in systems is attained by adding structure to the system (i.e., including partitions to avoid interaction and coupling). Indeed, as we discussed in the previous sections how entropy generation in business process instantiations is originating in the aggregation of information, the control and avoidance of this entropy generation seems to be situated in strictly partitioning the cost information structure. We now present a couple of principles or tentative guidelines in order to do so.

First, a number of *states* (i.e., "measuring points") should be incorporated in the design of the business process type such that relevant information aspects (e.g., "electricity consumption") are registered intermediately (i.e., regarding each task instantiation). The introduction of these states can be done in several ways: this might vary from asking actors who are executing the tasks to manually write down some of the needed information, to a software system automatically registering this information (e.g., by scanning batches or tracking the submission of work results by employees).

Second, these states should not be introduced in the design in an arbitrary way. Instead, a unique state should be introduced for each individual task (i.e., "information unit"), clearly separating the *information regarding each of these concerns*. While aggregated information can indeed be useful in several business situations for multiple stakeholders (cf. Section 4), it is important to record the information in this stage at most basic and fine-grained level. In case information at the aggregated level seems required at a later point in time, this information should be deductible from the elementary information gathered previously.

Third, taking the run-time perspective of our entropy analysis into account, a business process instantiation should be operating on only one information object and *each business process instance should be linked to the specific instance of the information object it is operating on*. This would allow ex-post analysis to take into account the specific characteristics of each individual information object (e.g., the size dimensions of the product to be manufactured or the specifics of a damage cost reimbursement request) and the influence this might have on the execution of a business process and its constituting tasks.

Finally, the *information regarding a task instantiation should be linked to the specific business process instantiation* it was embedded in (and hence, combined with guideline 3, linked to the information object it is operating on). When analyzing the origins of the entropy generation for our six discussed aggregation dimensions, one can easily find that the first two guidelines were violated in aggregation dimensions 2, 3, 5 and 6. The last two guidelines were violated in aggregation dimensions 4, 5 and 6. Consequently, consistently applying the four guidelines described above should enable a business process owner to end up in information aggregation dimension 1, exhibiting the lowest amount of entropy.

In fact, these guidelines can be directly derived from the Normalized Systems theory (NS) principles to control entropy generation during the run-time execution of software primitives [11]. Future research should hence be aimed at translating these general principles to more business-oriented and practical guidelines to more unambiguously control entropy. The application of modularity and NS reasoning to the organizational level, and the business process level in particular, is not that far-fetched. Indeed, the feasibility of applying the NS principles for attaining evolvability for the design-time implementation of business processes has been demonstrated previously [12].

4   Discussion

In our analysis, we focused on *business process analysis and optimizations in a managerial and operational context*, and motivated the need for fine-grained data (i.e., of aggregation dimension 1) based on entropy. Nevertheless, other business situations might require information at a different (i.e., more coarse-grained) aggregation level. For instance, for the purposes of (external) financial reporting or communication with several internal or external stakeholders (e.g., the board of directors) more aggregated information is obviously required. In these situations, there is no need to have information at the level of task and business process instantiations. Nevertheless, for certain purposes, fine-grained information is needed and a business should be systematically engineered in order to gather this data in a suitable way. Afterwards, this fine-grained information can still be easily summarized into more coarse-grained overviews.

Second, within this managerial and operational context, our analysis was focused at the *run-time complexity* of business process instantiations (i.e., once they are executed). The complexity generation in terms of entropy only becomes visible when data is generated through the execution of business process and task instances. While analysis at design-time can yield interesting research results as well, we remark that run-time analysis is researched less frequent in-depth.

4.1   Impact on Cost Accounting (Information) Systems in Practice

The relevance of fine-grained information of run-time processes can be observed in practical cost accounting designs. Consider two product varieties, which are produced using the industrial process presented in Section 3.1. Product A is a simple design, while product B is much more complex. While various tasks in the processes for both product types might be the same, the assembly task (i.e., task 4) is more expensive for product B. If aggregation dimension 4 is used, the distinction between the costs of the assembly tasks of both products cannot be made anymore. Instead, a general cost for the assembly task will be recorded, which is attributed evenly across products A and B. As a result, the cost of product A is overestimated, while the cost of product B is underestimated. In traditional cost accounting, which attributes costs based on volume-related measures (e.g., number of products produced), this is a valid approach. In literature, these approaches have indeed been described as insufficient when a large diversity of products is produced [13]. The entropy reasoning presented in this paper, might offer one way to understand the origin of this criticism.

An approach which claims to offer better management insights for complex business environments is Activity-Based Costing (ABC) [14]. ABC can be considered as a finer-grained way of performing cost accounting. More specifically, it focuses on attributing indirect costs to products. This is in line with our argument for the need for fine-grained information. An essential and initial step in the ABC approach is the identification of activities, which "are composed of the *aggregation* of units of work or tasks" [2, p. 342]. This definition suggests that ABC does not aim for aggregation dimension 1, since activities are defined as aggregations of tasks. Consequently, some critiques on ABC might be understood from the lack of fine-grained information as well.

The need for information at a fine-grained aggregation dimension can also impact the application portfolio of an organization. This was observed in a project in practice performed by one of the authors. In this project, an application was required for budget allocation. Such an application was already purchased from a software vendor, and was used to comply with legal financial reporting requirements. In the accounting application, costs could be attributed to a certain article. An article belongs to a certain activity, which is performed for a certain service. As a result, an overview of all costs for a service can be generated, as well as an overview for a certain type of activity (even for activities performed for different services). However, an article consists of certain products, which belong to certain domains. Budgets could be defined based on certain products as well. Because information concerning domains and products could not be stored in the accounting application, certain costs could not be attributed to the correct budget. Therefore, a finer-grained cost structure was registered in a spreadsheet document. As a result, every entry needed to be inserted twice, resulting in an increased employee workload, and duplication of data, which is detrimental from an information management viewpoint. This case illustrates why, even when information on a certain aggregation dimension is needed, requirements related to the information granularity need to be explored in-depth.

4.2 Limitations

The approach of this paper is explorative. As a result, several limitations should be taken into account. First, in this paper, we focused on the cost aspects of information units. It was argued that finer levels of granularity allow better diagnosis and traceability of issues related to these cost aspects. However, cost aspects are only a single dimension to be considered as a driver for creating fine-grained structures. In previous work, we identified various other dimensions which can determine such a structure, including throughput time or quality/output measures [9,10]. Further research is required to see if a fine-grained structure based on one dimension is identical to the structure when considering another dimension, or if different structures would be created. Second, an assumption in this paper was that the cost information related to an information unit could be observed uniformly. However, different costs may be involved in a certain business process task. For example, costs can be related to resources, employees, machine locking, etc. These different cost aspects have not been taken into account. Third, we used simplified examples in order to clarify our discussion. For example, each task in the business process was considered to cost around 5, which will not be the case in a realistic business process. Nevertheless, the same aggregation dimensions could be identified in that context. Detecting irregularities in processes would only become harder, since deviations in relation to the expected mean would need to be incorporated in the calculation. Fourth, we considered the instantiation space of a single business process. In reality, multiple business processes could be in scope when diagnosing a certain phenomenon. This would result in an instantiation space containing business process instances of different business process types and more complex aggregation dimensions could be defined for combinations of instantiations of these different business processes. In these cases, entropy would increase even more. Finally, as our rationale is primarily based on theoretical reasoning, a set of case studies should be performed in future research to further validate our claimed hypotheses. Also, the trade-off in practice between more fine-grained cost information (exhibiting less entropy) and the actual costs for assembling this information might constitute an interesting avenue for further investigation.

5   Conclusion

In this paper, we explored how entropy is generated during the run-time execution of business processes. We focused on entropy generated through information aggregation, which we described using six different aggregation dimensions. Moreover, we discussed in-depth how this entropy generation impacts cost accounting aspects, both in the design of cost accounting systems and their supporting information systems.

Acknowledgment. P.D.B. is supported by a Research Grant of the Agency for Innovation by Science and Technology in Flanders (IWT).

References

1. Kaplan, R.S., Anderson, S.R.: Time-driven activity-based costing. Harvard Business Review 82(11), 131–138 (2004)
2. Drury, C.: Management and Cost Accounting. South-Western (2007)
3. Lev, B.: The aggregation problem in financial statements: An informational approach. Journal of Accounting Research 6(2), 247–261 (1968)
4. Ronen, J., Falk, G.: Accounting aggregation and the entropy measure: An experimental approach. The Accounting Review 48(4), 696–717 (1973)
5. Abdel-Khalik, A.R.: The entropy law, accounting data, and relevance to decision-making. The Accounting Review 49(2), 271–283 (1974)
6. Boltzmann, L.: Lectures on Gas Theory. Dover Publications (1995)
7. Wikipedia: Entropy (2013), http://en.wikipedia.org/wiki/Entropy
8. Jung, J.Y., Chin, C.H., Cardoso, J.: An entropy-based uncertainty measure of process models. Information Processing Letters 111(3), 135–141 (2011)
9. De Bruyn, P., Huysmans, P., Oorts, G., Mannaert, H.: On the applicability of the notion of entropy for business process analysis. In: Proceedings of the Second International Symposium on Business Modeling and Sofware Design (BMSD), pp. 93–99 (2012)
10. De Bruyn, P., Mannaert, H.: On the generalization of normalized systems concepts to the analysis and design of modules in systems and enterprise engineering. International Journal on Advances in Systems and Measurements 5(3&4), 216–232 (2012)
11. Mannaert, H., De Bruyn, P., Verelst, J.: Exploring entropy in software systems: towards a precise definition and design rules. In: Proceedings of the Seventh International Conference on Systems (ICONS), pp. 93–99 (2012)
12. Van Nuffel, D.: Towards designing modular and evolvable business processes. PhD thesis, University of Antwerp (2011)
13. Atkinson, A., Banker, R., Kaplan, R.: Management Accounting. The Robert S. Kaplan Series in Management Accounting. Prentice Hall (2001)
14. Kaplan, R.S., Bruns, W.: Accounting and Management: A Field Study Perspective. Harvard Business School Press (1987)

Author Index